Critical Issues in Tourism Co-Creation

Critical Issues in Tourism Co-Creatio

Co-creation is fast becoming a buzz word in tourism. Traditional approaches operators and suppliers produce goods and services which are consumed by measured in economic terms. Co-creation challenges these assumptions, arguing co-create value together and that this value is more diverse than just economic v media, ratings and review tools and e-commerce are facilitating the creation of dive for driving innovation in, for example, new business models such as the collabora emotional, reputational and other kinds of value may also be produced, and a wide r cers or consumers, might also benefit from the value co-creation process.

This edited volume seeks to go beyond the dominant business/management/ma the co-creation of market value and innovation, to excavate complex and critical epis engaging authors from both the academy and beyond, it explores the rich historical lina porary practices.

The chapters in this book were originally published in *Tourism Recreation Research*.

Giang Thi Phi is currently a researcher at the Hospitality and Tourism Institute, Faculty of So Duy Tan University, Danang, Vietnam. Giang holds a PhD in Tourism Management from Gri researches in the areas of event and tourism, social entrepreneurship, innovation and sustain Giang organised the first TEFI Walking Workshop in Vietnam to explore pathways for a Sch Tourism & Social Entrepreneurship.

Dianne Dredge is Director of The Tourism CoLab, professor, entrepreneur and an internationa leader. She works with business, communities and governments to co-create solutions for more su regenerative tourism futures. Dianne is an environmental planner, with higher degrees in tourism, and organisational change.

Critical Issues in Tourism Co-Creation

Edited by
Giang Thi Phi and Dianne Dredge

Routledge
Taylor & Francis Group
LONDON AND NEW YORK

First published 2022
by Routledge
2 Park Square, Milton Park, Abingdon, Oxon OX14 4RN

and by Routledge
605 Third Avenue, New York, NY 10158

Routledge is an imprint of the Taylor & Francis Group, an informa business

British Library Cataloguing in Publication Data
A catalogue record for this book is available from the British Library

ISBN: 978-0-367-76179-0 (hbk)
ISBN: 978-0-367-76180-6 (pbk)
ISBN: 978-1-003-16583-5 (ebk)

Typeset in Myriad Pro
by Newgen Publishing UK

Publisher's Note
The publisher accepts responsibility for any inconsistencies that may have arisen during the conversion of this book from journal articles to book chapters, namely the inclusion of journal terminology.

Contents

Citation Information

The following chapters, except Chapter 8, were originally published in *Tourism Recreation Research*, volume 44, issue 3 (September 2019). Chapter 8 was originally published in volume 40, issue 3 of the same journal. When citing this material, please use the original page numbering for each article, as follows:

Chapter 1
Collaborative tourism-making: an interdisciplinary review of co-creation and a future research agenda
Giang T. Phi and Dianne Dredge
Tourism Recreation Research, volume 44, issue 3 (September 2019), pp. 284–299

Chapter 2
Wildlife tourism through the co-creation lens
Giovanna Bertella, M. Fumagalli and V. Williams-Grey
Tourism Recreation Research, volume 44, issue 3 (September 2019), pp. 300–310

Chapter 3
Co-creating knowledge in tourism research using the Ketso method
Yana Wengel, Alison McIntosh and Cheryl Cockburn-Wootten
Tourism Recreation Research, volume 44, issue 3 (September 2019), pp. 311–322

Chapter 4
Co-creating an integrated curriculum alongside community partners: a creative analytic approach
Karla Boluk, Meghan Muldoon and Corey Johnson
Tourism Recreation Research, volume 44, issue 3 (September 2019), pp. 323–336

Chapter 5
Student living labs as innovation arenas for sustainable tourism
Eva Maria Jernsand
Tourism Recreation Research, volume 44, issue 3 (September 2019), pp. 337–347

Chapter 6
'Dig where you stand': values-based co-creation through improvisation
José-Carlos García-Rosell, Minni Haanpää and Jenny Janhunen
Tourism Recreation Research, volume 44, issue 3 (September 2019), pp. 348–358

Chapter 7
The co-creation of diverse values and paradigms in small values-based tourism firms
Lucia Tomassini
Tourism Recreation Research, volume 44, issue 3 (September 2019), pp. 359–369

Chapter 8

The changing face of the tour guide: one-way communicator to choreographer to co-creator of the tourist experience
Betty Weiler and Rosemary Black
Tourism Recreation Research, volume 40, issue 3 (2015), pp. 364–378

Notes on Contributors

Giovanna Bertella, School of Business and Economics, UiT The Arctic University of Norway, Tromsø, Norway.

Rosemary Black, School of Environmental Sciences, Charles Sturt University, Albury, NSW, Australia.

Karla Boluk, University of Waterloo, Waterloo, Canada.

Cheryl Cockburn-Wootten, Waikato Management School, The University of Waikato, Hamilton, New Zealand.

Dianne Dredge, Department of Service Management and Service Studies, Lund University, Lund, Sweden; The Tourism CoLab, Brisbane, QLD, Australia.

M. Fumagalli, Tethys Research Institute, Milano, Italy.

José-Carlos García-Rosell, Faculty of Social Sciences, University of Lapland, Rovaniemi, Finland.

Minni Haanpää, Faculty of Social Sciences, University of Lapland, Rovaniemi, Finland.

Jenny Janhunen, Seinäjoki University of Applied Sciences, Seinäjoki, Finland.

Eva Maria Jernsand, Department of Business Administration, University of Gothenburg, Gothenburg, Sweden.

Corey Johnson, University of Waterloo, Waterloo, Canada.

Alison McIntosh, School of Hospitality and Tourism, Auckland University of Technology, Auckland, New Zealand.

Meghan Muldoon, HNU-ASU, Haikou City, Hainan, People's Republic of China.

Giang T. Phi, Tourism Research Network, The Faculty of Humanities, Aalborg University, Aalborg, Denmark.

Lucia Tomassini, School of Events, Tourism and Hospitality Management, Leeds Beckett University, Leeds, UK.

Betty Weiler, School of Business & Tourism, Southern Cross University, Coolangatta, Australia.

Yana Wengel, Strathclyde Business School, University of Strathclyde, Glasgow, United Kingdom.

V. Williams-Grey, Whale and Dolphin Conservation, Plymouth, MA, USA.

Introduction: Critical issues in tourism co-creation

Giang T. Phi and Dianne Dredge

Historical antecedents, definitional issues and disciplinary influences

'Co-creation' is a relatively new label for a range of collaborative practices, many of which have been discussed at length in tourism scholarship (Bramwell & Lane, 2011; Jamal & Getz, 1999). In its most basic conceptualisation, and for the purpose of this book, co-creation refers to collaborative, participatory practices whereby multiple, and often diverse, actors help to produce something of value together. Each participant sees the benefits of participating and co-creating a solution, a product, an experience or other outcome differently. Moreover, the value that they obtain could also be vastly different and take on different tangible and intangible forms.

What we think is so interesting about the notion of co-creation, and that we seek to explore in this book, is that co-creation simultaneously shines a light on: (i) the processes of collaboration through which something is co-created, (ii) the value that is produced and the different forms that value can take, and (iii) the distribution of value across different groups of actors, who may be direct or indirectly involved. In essence, the concept of co-creation pulls apart the production–consumption binary that has remained unquestioned for so long, and it helps us understand why actors are motivated to collaborate and work together. This exploration flows naturally from earlier investigations in networked governance (Dredge, 2006), in collaborative economy (Dredge & Gyimóthy, 2016) and in diverse economies (Cave & Dredge, 2020).

Basing their arguments on simple database searches focusing on a narrow set of terms, some researchers argue that 'co-creation' has only just started to emerge in the last decades. Without taking stock of the rich historical lines of research and identifying its lineage within other fields of study including planning, community engagement, sociology, international development and so on, some have argued that co-creation is innovative – a new way to undertake research and engage with actors. In the process, binaries are broken down between researcher and subject, and there is a blurring of distinctions between problem and action, and process and outcome. Co-creation in this sense captures the notion of 'co-labour-ation'.

Other researchers (e.g., Bason, 2010) associate co-creation with user-driven innovation, demand-driven innovation, customer-driven innovation, human-centred design, interaction design and so on. Usurped into business and management studies, this discourse tends to be closely associated with ideals such as economic innovation, efficiency, timeliness, effective participation that are generally associated with neoliberal values such as shared responsibility, public–private partnerships and capacity building (Vargo & Lusch, 2004; Vargo & Lusch, 2008; Zwick, Bonsu, & Darmody, 2008; Terblanche, 2014). It was this uneven and blinkered approach to the emergent term 'co-creation', and the lack of acknowledgement of the interdisciplinary origins of diverse applications of collaboration in tourism, that provided the initial inspiration for this book.

We have sought to excavate and acknowledge the deeper historical antecedents, and how this emergent term 'co-creation' might also be understood as a re-interpretation of long standing values associated with community empowerment, collaboration, the (re) democratisation of planning and policy processes (e.g., see Bosman & Dredge, 2015; Dredge & Hales, 2012; Phi, Whitford, & Dredge, 2017). The term 'co-creation' also makes use of long-established discussions about knowledge production and dynamics, where different types of knowledge (e.g., explicit (written, encoded, etc.) and tacit (e.g., embrained, encultured, embodied)) co-exist, coalesce and feed off one another to be valued by stakeholders in different ways (see Nonaka & Takeuchi, 1995; Dredge & Jenkins, 2011). In essence then, our interest was in bringing together diverse interpretations of co-creation and collaboration in tourism and to explore how these have influenced our understanding of tourism governance, planning, knowledge dynamics, production and consumption, community empowerment, international development and so on. We wanted to move the discussion beyond service-dominant logic, which is where co-creation in tourism has been so narrowly focused for the last few years.

Co-creation and the human–nature interface

Our position is that the concept of co-creation – the working together to produce or create something of value – is much older dimension of the co-operative and collaborative human condition. Co-creation, collaboration, shared production, partnerships and co-operation similarly capture the idea that value is produced by working together. Yet in taking this approach we should also be aware that the value produced may have positive and/or negative value, and the determination of that value is different for different actors. Moreover, the co-creation process might be organic and emergent, such as an experimental action, or it might be carefully anticipated, designed and produced, such as in the case of a tourism planning framework (see Dredge, Ford, & Whitford, 2011).

But there are also exciting advances that have emerged in the concept of co-creation, and that have prompted us to think differently, that we need to acknowledge. In particular, an important development emerging from co-creation discussions has been to shine the light on what is produced from co-labour-ation. The outcome or product could be tangible, such as a policy or agreed action, or it could be intangible such as social, economic or cultural value, trust, knowledge or understanding.

Co-creation can also be transformative, producing new hybrid actors – human and non-human (Haraway, 2015). As a result, discussions of co-creation prompt us to reorder our very human perspective on the world and to try to understand from the perspective of Nature, for example (Chakraborty, 2019). To illustrate, in nature-based destinations, travellers, the tourism industry and Nature co-create visitor experiences. And, if we take this view, then the question emerges as to what value Nature gets out of its participation. In essence, it helps us to look at tourism differently, to identify alternative viewpoints and to develop innovative approaches.

The work of Harraway and others suggest that what is produced might also create new hybrid things. With their own influence and agency, the best-known example of this new hybrid actor is the Great Pacific Gyre, a complex actor in its own right, produced through complex physical, economic, chemical, environmental and human interactions (Ballantyne & Varey, 2006). What is clear from this discussion is that co-creation is that we also need to extend our thinking about co-creation into the realm of interactions between human and non-human things. These developments highlight that co-creation also has moral and ethical dimensions; we must engage with the diverse and often complex outcomes, whether they are positive or negative or both; and what we value, how we value and who values what is produced. In taking this approach, we must also remain alert to the way in which the emergence of new philosophical directions in the humanities and social sciences, such as post-structuralism and post-humanism, are adding to and deepening our understandings of co-creation as a process of becoming together. We have tried to embrace these new directions as well as more established investigations of co-creation in tourism in this book.

Contributions in this book

This edited book brought together a range of contributions not only from diverse geographical, developmental, scalar and temporal contexts but also they were co-created by diverse authors from both the academy and beyond (e.g., business administrator, biologist, policy manager, social entrepreneur).

Phi and Dredge unpack the historical roots and characteristics of co-creation from seven threads of scholarship, which provide a broader umbrella to expand the understanding of co-creation in tourism. Among these threads, co-creation from a post-human perspective is explored further by Bertella, Fumagalli and Williams-Grey in their research on wildlife tourism. Utilising the case of swim-with-dolphins tour, these authors suggest that wild animals can potentially become the key actor in co-creating knowledge relevant for wildlife tourism management. Their research also emphasises the role of empathy between human and non-human actors as the starting point for effective co-creation processes beyond a human-centred approach.

Three other chapters approach co-creation using participatory, action-oriented research and problem-solving methodologies. Wengel, McIntosh and Cockburn-Wootten suggest that the Ketso method can be an important tool to add to the suit of existing collaborative methodologies. Along the same line, Boluk, Muldoon and Johnson critically explore Integrated Curriculum Design as a creative pedagogical tool to co-instruct and co-learn with diverse community partners. Jernsand examines co-creation in 'student living labs', demonstrating the value of open innovation and experimentation in action research. Though diverse in the approaches, methodologies and tools used, these chapters clearly demonstrate that binaries are breaking down, and concurrently, the blurring of lines between researcher and subject, the problem and solution, process and action.

The last three chapters approach co-creation through the values of small tourism firms and tour guides. Garcia-Rosell, Haanpää and Janhunen offer a more comprehensive understanding of value co-creation processes in small tourism firms by drawing upon cultural marketing and organisational improvisation. Tomassini further expands the topic by highlighting the co-creation of an ethical vision of non-commercially oriented tourism firms, and their contribution to co-creating alternative values and paradigms in tourism with regard to development, growth, citizenship and entrepreneurship. Finally, Weiler and Black explore the evolving roles of tour guides in co-creating tourist experiences. They highlight the increased importance of tour guides as 'facilitator of the tourists' inner journey', which can be transformative, life changing and can lead to further positive values being created after the tourists returned home.

Three major themes emerge from this book. First, the chapters contribute to challenge the idea that useful knowledge is only produced by scientific communities. Rather co-creation is conceived as a metaphor for a range of established research practices and approaches that bridge the gap between science and society, where the researchers are positioned as facilitators and co-creators rather than lone experts (see e.g., Wengel, McIntosh & Cockburn-Wootten – chapter 3). Second, co-creation in tourism can generate positive values for diverse range of actors involved, including both human and non-human actors. These positive values can vary widely from individual level (e.g., financial stability, cultural understanding, happiness and well-being), organisational level (e.g., enhanced competitiveness, efficiency and resilience) to societal/ecosystem level (e.g., social cohesion, pride, social justice and sustainability). Third, through taking a critical approach towards co-creation, this book also cautions readers that co-creation processes may not always produce positive values. Insufficient awareness of the relations between actors involved in the co-creation process (e.g., equality, emotions, power) may be breeding grounds for value co-destruction, instead of value co-creation, in tourism context (e.g., Phi & Dredge – chapter 1; Garcia-Rosell, Haanpää & Janhunen – chapter 6). Thus, in excavating co-creation from historical, multi-disciplinary and multi-actor perspectives, this book seeks to lay the foundations for a more comprehensive examination of value co-creation in future tourism research, in order to fully realise tourism's potential as a powerful co-creative social force for good.

Finally, in the journey of developing this book, we also acknowledge a certain cynicism among those that might see the term 'co-creation' as part of a language game played within academia, where labelling and concept claiming can be used by researchers to boost citations, improve metrics and consolidate one's scholarly reputation. In the process, historical threads of theoretical development are discarded and scholarly publishing falls into faddishness. It is our ambition in this book to transcend such practices, and to recognise that working together to co-produce value, collaborative understanding, and joint outputs is fundamental to addressing the complex challenges we face. We encourage readers to engage critically with the term 'co-creation', to recognise related elements such as co-design and co-production, and to acknowledge not only the contributions stemming from diverse disciplines, but also the innovative future inherent in collaborative working. The Anthropocene demands that we de-centre our human perspective on collaboration, to exercise empathy and to acknowledge the rights of Nature. Co-creation has an enormous contribution to make in this regard, because it implores us to think about the co-design, co-creation and co-production of tourism *with* Nature, and not simply as based on, or exploiting, Nature.

References

Ballantyne, D., & Varey, R. J. (2006). Creating value-in-use through marketing interactions: The exchange logic of relating, communicating, and knowing. *Marketing Theory*, *6*(3), 335–348. doi: 10.1177/1470593106066795

Bason, C. (2010). Design thinking in government. In C. Bason (Ed.), *Leading public sector innovation: Co-creating for a better society* (pp.135–149). Bristol: Policy Press.

Bosman, C., & Dredge, D. (2015). *Teaching about tourism in a post-disciplinary planning context*. In D. Dredge, D. Airey, & M.J. Gross (Eds.), *The Routledge handbook of tourism and hospitality education* (pp. 265–278). Abingdon: Routledge.

Bramwell, B., & Lane, B. (2011). *Critical perspectives on governance and sustainability*. London: Routledge.

Cave, D., & Dredge, D. (2020). *Reworking tourism: Diverse economies in a changing world*. New York: Routledge.

Chakraborty, A. (2019). Does Nature matter? Arguing for a biophysical turn in the ecotourism narrative. *Journal of Ecotourism*, doi: 10.1080/14724049.2019.1584201

Dredge, D. (2006). Policy networks and the local organisation of tourism. *Tourism Management*, *27*(2), 269–280.

Dredge, D. & Gyimóthy, S. (2017). *Collaborative economy & tourism: Policy, prospects and perspectives*. Springer: Cham, Switzerland.

Dredge, D., & Hales, R. (2012). Community case study research. In L. Dwyer, A. Gill, & N. Seetaram (Eds.), *Handbook of research methods in tourism: Quantitative and qualitative approaches* (pp. 417–437). Cheltenham: Edward Elgar.

Dredge, D., & Jenkins, J. (2011). New spaces of tourism policy and planning. In D. Dredge, & J. Jenkins (Eds.), *Stories of practice: Tourism planning and policy* (pp. 1–12). Surrey: Ashgate.

Dredge, D., Ford, E. -J., & Whitford, M. (2011). The managing local tourism master class: Building sustainable management practices across local government divides. *Tourism and Hospitality Research*, *11*(2), 101–116. doi: 10.1057/thr.2010.064

Haraway, D. (2015). Anthropocene, Capitalocene, Chthulhucene. Donna Haraway in conversation with Martha Kenney. In H. Davis, & E. Turpin (Eds.), *Art in the Anthropocene: Encounters among aesthetics, politics, environments and epistemologies* (pp. 255–269). London: Open Humanities.

Jamal, T., & Getz, D. (1999). Community roundtables for tourism-related conflicts: The dialectics of consensus and process structures. *Journal of Sustainable Tourism*, *7*(3–4), 290–313. doi: 10.1080/09669589908667341

Nonaka, I., & Takeuchi, H. (1995). *The knowledge-creating company: How Japanese companies create the dynamics of innovation*. Oxford: Oxford University Press.

Phi, G. T., Whitford, M., & Dredge, D. (2017). Knowledge dynamics in the tourism - social entrepreneurship nexus. In P. Sheldon, & R. Daniele (Eds.), *Tourism and Social Entrepreneurship: Philosophy and Practice* (pp. 155–172). Cham: Springer.

Terblanche, N. S. (2014). Some theoretical perspectives of co-creation and co-production of value by customers. *Acta Commercii*, *14*(2), Art. #237, 8 pages. doi: 10.4102/ac.v14i2.237

Vargo, S. L., & Lusch, R. F. (2004). Evolving to a new dominant logic for marketing. *Journal of Marketing*, *68*(1), 1–17. doi: 10.1509/jmkg.68.1.1.24036

Vargo, S. L., & Lusch, R. F. (2008). Service-dominant logic: Continuing the evolution. *Journal of the Academy of Marketing Science*, *36*(1), 1–10. doi: 10.1007/s11747-007-0069-6

Zwick, D., Bonsu, S. K., & Darmody, A. (2008). Putting consumers to work: "Co-creation" and new marketing govern-mentality. *Journal of Consumer Culture*, *8*(2), 163–196. doi: 10.1177/1469540508090089

Collaborative tourism-making: an interdisciplinary review of co-creation and a future research agenda

Giang T. Phi and Dianne Dredge

ABSTRACT

For some time, tourism researchers have sought to examine and theorise types of collaborative exchange and the characteristics of relational work in tourism. Different ontological and epistemological framings, and associated language games have contributed to a fragmented body of knowledge. In this paper, we argue that the new term 'co-creation' is part of this language game, and efforts to date have not linked co-creation to the broader and deeper currents of theory building that have come before. We thus place co-creation within its wider context by, firstly, building a meta-narrative review of the literature that draws together a number of disparate disciplinary-inspired lines of thinking, and secondly, by identifying and extending key concepts of co-creation and its logics to tourism. We trace seven threads of scholarship that demonstrate the ideas and values associated with co-creation have diverse historical roots. Using a meta-narrative approach, we unpack the characteristics of co-creation from different disciplinary lenses, directing attention to issues beyond service-dominant logic approaches towards wider issues of participation, inclusion, power, responsibility, and value. In the process, we contribute to a new and fresh appreciation of value co-creation in tourism literature, along with a nine-point agenda that suggest directions for future research and practice.

Introduction

In 2016, Copenhagen's destination marketing agency, Wonderful Copenhagen, declared 'the end of tourism as we know it':

> … it's time to welcome the new traveler – the temporary local, seeking not the perfect picture to take home but the personal connection to an instantly shared experience based on interest, relations and authenticity. In other words, we need to set course towards a future without tourism as we know it. Because by doing so, we can start to focus on something much more interesting: A future of hosts, guests and a shared experience of localhood. (http://localhood.wonderfulcopenhagen.dk/)

This declaration illustrates the rise of co-creation as an ideological force in tourism, and how it is shaping ideas about what is value, where value is created, who creates it, and who is responsible for its creation (see e.g. Campos, Mendes, do Valle, & Scott, 2016). In declaring their shifting role from an agency focused primarily on marketing to a broader, more collaborative placemaking and marketing role, the DMO argued for the adoption of 'localhood' (Wonderful Copenhagen, 2017). The localhood, they argued, encouraged destination actors to think of visitors as temporary locals rather than tourists. In the context of rising concerns about overtourism across Europe, the localhood was also aimed at breaking down tensions inherent in traditional terminologies, such as tourists and residents, locals and visitors, home and away, destination and residential areas, and so on. The localhood was thus framed as a place collaboratively created through diverse encounters between visitors and residents. It also marked a shift in thinking about the role (and power) of the DMO from leader-in-charge towards being a facilitator of visitor experiences in a diverse city-scape (Čorak & Živoder, 2017). So, in addition to marketing the City of Copenhagen to the outside world, the DMO also turned its attention towards better understanding and facilitating successful visitor experiences and positive outcomes for the city's temporary and permanent inhabitants. Across the world, Wonderful Copenhagen's declaration was posted, reposted, tweeted and retweeted on social media. Comments amounted to a collective celebration that a major and innovative DMO like Wonderful Copenhagen was acknowledging that tourism is much more than visitor numbers and expenditure, that collaboration across policy sectors was important, and that the blurring of categories like 'resident' and 'visitors' could open up innovation and reframe how we think of tourism.

The above example of Copenhagen illustrates a shift in thinking about where value is created and the type of value that is created in tourism (Wonderful Copenhagen, 2017). Value is not just created within the tourism sector by operators and the DMOs but is also generated through interactions and exchanges between a wide range of human and non-human actors both inside and outside the destination (Buonincontri, Morvillo, Okumus, & van Niekerk, 2017; Jensen & Prebensen, 2015). Further, the DMO is not solely responsible for generating, nurturing, and managing the value created, but it is a collaborative responsibility, and success rests on a range of factors including collaboration, synchronicity, shared value, trust, and so on (Cabiddu, Lui, & Piccol, 2013). Indeed, the creation of shared value in tourism is receiving growing attention from a wide range of researchers in marketing, governance, product development, innovation systems, to name a few areas (Lee, Olson & Trimi, 2012).

Our point of departure for this paper is that tourism research has, for some time, sought to examine and theorise types of collaborative exchange and the characteristics of relational work in tourism (Bramwell & Lane, 2000; Hall, 1999; Jamal & Getz, 1995). However, different ontological and epistemological framings, and their associated language games have contributed to a fragmented body of knowledge. We posit that the use of the new term 'co-creation' is part of this language game, and tends not to acknowledge the broader and deeper currents of theory building that have come before. Put simply, tourism co-creation is increasingly used as a 'buzz' word, often adopted on a rather superficial level, and without consideration of the history and the broader development of 'co-creation' literature in other disciplines. We also need to acknowledge that co-design, co-creation, and co-production are different forms of collaborative exchange, and that it is important not to simply adopt co-creation as an all-encompassing term. Our own stance is that current attention on tourism co-creation reflects the relational turn in the social sciences, an ontological shift from a predominantly rational scientific view of the world to a socially constructed and interdependent world (Powell & Dépelteau, 2013). But co-creation is not a cohesive theoretical project; it is a metaphor prone to abstraction, and provides little theoretical direction for the development of tourism studies as a field. This turn can be traced back through a linage of scholars as diverse as Karl Marx, Georg Simmel, Pierre Bourdieu and Bruno Latour. Globalisation, digitalisation, and subsequent recognition that we need to rethink spatio–temporal–material relations have fed recent attention to this turn towards collaborative creation of value and co-production (Eacott, 2018).

Our aim in this paper is two-fold. Firstly, we seek to critically examine and extend the notion of collaborative tourism making by undertaking a meta-narrative analysis of co-creation. Secondly, we seek to identify and extend key concepts of co-creation, and in the process build and understanding of co-creation as something relevant to tourism researchers, and not just limited to those working with service-dominant logic. Our starting point is that the very act of exchange, the collaborative creation and co-production of something, such as an experience, a marketing message, a product, a service, and so on, is what constitutes tourism. We propose that developing a broader perspective on co-creation can transform how we understand and make sense of tourism and its transformative effects on people, places, and the planet. In much the same way that mobilities studies have provided a new theoretical lens to understand tourism, we believe that taking a more systematic approach to tourism co-creation, and drawing together the diverse theoretical tangents of co-creation can provide a powerful lens to understand tourism better.

In this paper then, we seek to place co-creation within its wider context by, firstly, building a meta-narrative that draws together a number of disparate disciplinary-inspired lines of thinking, and secondly, by identifying and extending key concepts of co-creation and its logics to tourism. We follow the concept of co-creation and its rise within the tourism literature, while also acknowledging the various ontological and epistemological roots within other disciplines and fields of study that have shaped how it is framed and applied in tourism research. Hence, we have deliberately decided not to provide a definition of co-creation here at the beginning of the paper, but to discuss its meaning in later sections after having reviewed the literature. Through this process, the paper seeks to contribute a fresh appreciation and more comprehensive understanding of value co-creation in tourism literature, along with the proposal of a nine-point agenda for future research and practices.

Approach: a meta-narrative analysis of co-creation

There has already been a significant amount of work done in theorising co-creation from various disciplinary perspectives, with Table 1 identifying a number of systematic literature reviews completed to date.

These reviews have predominantly been generated from management and service studies. While one systematic review on co-creation in tourism was identified

Table 1. Systematic reviews of co-creation.

Author	Discipline/Subject area	Keywords	Purpose of review	Contribution
Velamuri, Neyer and Möslein (2011)	Organisational management	Hybrid value creation, customer solutions, product service systems, integrated solutions, servitization, complex product systems	A systematic review of 169 publications on hybrid value creation (i.e. combining product and service offering) in commercial firms	Identified dynamic resources and capabilities as being important to co-creation
Galvagno and Dalli, (2014)	Service management	Value, co-creation, service science, customer participation, service-dominant logic	Systematic review of 72 influential publications (–2014) (utilising co-citation) to explore the past, present and future state of value co-creation in service management	Identified three key theoretical perspectives on value co-creation: service science, innovation and technology management, and marketing and consumer research
Voorberg, Bekkers and Tummers (2015)	Public management	Co-creation, co-production, public sector innovation, social innovation	A systematic review of 122 publications (1987–2013) of co-creation in the context of social and public innovation	Co-creation is associated with co-production and active citizen involvement, terms frequently used in public management literature
Campos et al. (2016)	Tourism management	Co-creation tourism experience; active participation; interaction; tourist psychology	A systematic review of 50 publications (2008–2015) on the co-creation of tourist experience	Identified key dimensions of co-creation for creating tourist experience
Mandrella, Zander, and Kolbe (2016)	System sciences/ IT management	Value co-creation, business value of information technology, inter-organizational networks	A systematic review of 45 publications (2000–2015) on IT-based value co-creation at the inter-organisational level	Authors suggest a framework to understand value in co-creation, which analyses IT-based value at the network, relation-specific, and firm levels. Value is seen as manifest itself in performance, processes, or intangible (e.g. loyalty) dimensions
Alves, Fernandes, and Raposo (2016)	Business and management	Value co-creation, innovation, consumer experience, market relations	Clarifies the field of application and study of co-creation	Authors conclude that field of knowledge has developed after 2007, Four clusters of research identified: business innovation, new porducts and services; service provision; and relational marketing
Tekic and Willoughby (2018)	Organisational management	Co-creation; open innovation; innovation management; collaborative innovation	Two-stage systematic review (broad and in-depth) of 77 publication (–2017) on co-creation in the innovation management literature	Authors put forward a simple and clear definition of co-creation, comparing and contrasting with open innovation

(Campos et al., 2016), its focus is limited to an examination of co-creation as a component of destination competitiveness and tourists' roles in creating commercial visitor experiences. Furthermore, none of these systematic reviews attempt to bridge disciplinary boundaries or consider the diverse relational roots of co-creation. As a result, the words used to perform bibliographic searches in these systematic reviews were often limited to the terms 'co-creation' and closely related semantic expressions such as 'co-production', 'customer participation' and 'active involvement', together with subject area terms such as 'tourism' or 'service'. These systematic reviews also utilise formal databases, which favour serial publications with an ISSN (International Standard Serial Numbers), and subsequently exclude the significant body of grey literature.

We posit that the development of a meta-narrative understanding of tourism co-creation across disciplinary divides would benefit tourism studies by providing a fresh, novel conceptualisation. Over 30 years ago, Normann and Ramirez (1993) observed that thinking about value creation was locked in an industrial economy mindset, a criticism that still appears relevant. Saarijärvi, Kannan, and Kuusela (2013) further observe that without systematic and analytical clarification, the utility of the concept of co-creation is diminished. These authors argue that dismantling 'value', 'co', and 'creation' are key to discovering the multifaceted nature of co-creation, a point which we also agree.

So, returning to our approach in this paper, whilst a thematic analysis on the above systematic reviews would bring together some major themes related to co-creation, these existing reviews are limited to business and management fields, and it is unlikely that a thematic analysis would reveal additional insights that would provide a useful, novel, or fresh understanding. The point is that the relational work through which value is co-produced can be expressed very differently depending on the discipline or field of study, so a systematic review of literature based on 'co-creation' and related terms will yield narrow results. The difficulty of identifying appropriate search terms across a fragmented body of knowledge in different disciplines was one challenge, but we were also seeking an understanding of the interactive, collaborative, relational, and value-making dimensions of co-creation. We posit that, in addition to the traditional economic, business, and management foci, the relational work of co-creation produces social, political and other kinds of value that are not configured in the above reviews. This observation demanded a closer reading of the literature, knowledge of interdisciplinary connections, and a deeper abductive approach to theory building, features that are inherent in a meta-narrative review (Fleury-Vilatte & Philippe Hert, 2003; Snilstveit, Oliver, & Vojtkova, 2012).

A 'meta-narrative review' examines how a particular research area has unfolded over time, how it has shaped the kinds of questions being asked, and the influence these historical antecedents have played on the dominant methods being employed. In other words, 'they examine the range of approaches to studying an issue, interpret and create an account of the development of these separate narratives and then create an overarching meta-narrative summary' (Gough, 2013, p. 2). The challenge of the meta-narrative review is that methodologies can vary widely due to diversity in ideological assumptions, general methodological approaches, specific case studies methods, that are present within the particular streams of literature making up the wider body of research. Gough (2013) identified two broad streams of meta-review: (1) 'aggregate reviews' that aim to aggregate findings within a predetermined conceptual framework, and (2) 'configuring reviews' that aim to configure, interpret and arrange theories and concepts by employing iterative methods and emergent concepts.

We adopt a configuring meta-narrative approach in this paper, focusing on how tourism co-creation has been researched with particular emphasis on the ideas, data and methods used, rather than synthesising the findings of the research. So, on the one hand, we utilise partially explicit knowledge in both existing systematic and narrative reviews to configure overarching themes. These reviews were identified by searching for 'co-creation' OR 'value co-creation' AND 'review' in four major research databases: Scopus, Web of Science, ScienceDirect and Proquest. On the other hand, we supplement this with our own tacit knowledge (accumulated over 40 years of combined experience) from allied disciplines and fields of study (urban and environmental planning, sociology of leisure, business, economic development, politics, development studies, policy and governance) to critically question and to unearth missing perspectives and knowledge not present in the existing systematic and narrative reviews. In adopting this approach, we do not seek to produce linear continuous historiography, but rather adopt a post-structural archaeological approach where different knowledge contributions co-exist and overlap (c.f. Foucault, 1969; Scheurich, 1994).

Based on both the existing reviews and tacit knowledge, we identified the key terminologies in co-creation in various disciplines (Table 2), which then informed the search and inclusion of relevant references and discussions in this meta-narrative review.

The Table is by no means an exhaustive list of terminologies, but it serves to demonstrate the deeper and diverse roots of co-creation across diverse disciplines, and also the common elements in these discussions. Potential value of this Table then, is to facilitate boundary spanning scholars, enabling them to continue to evolve their interdisciplinary thinking in new contexts.

An archeology of knowledge in tourism co-creation

According to Kuhn (1970), the evolution, maturation, and uptake of knowledge occur in paradigm shifts defined as 'universally recognised scientific achievements that for a time provide model problems and solutions to a community of practitioners' (p. xiii). In an attempt to soften the perceived rigidity of paradigms, Lipman (1991, 2003) and Paul (2011) have argued that knowledge comes in waves. In the first wave, a new idea is often enthusiastically embraced, supported and reinforced by researchers. A second wave occurs some time later as cognitive processes, reflective scepticism, reasoning, judgment, and argumentation develop. The absorption of knowledge is influenced by the social worlds inhabited by different tourism actors, and readiness for learning and reflecting provided within these different contexts.

In practice, this is illustrated in the different social worlds in which tourism actors circulate, reinforced by dense social ties with their own kind, that serve to limit opportunities for communication and knowledge sharing (Phi, Dredge, & Whitford, 2014). Over time, judgement and over-simplified characterisations of those in other social worlds reduce discursive engagement and the collaborative processing of insights and knowledge. Knowledge brokers, such as consultants, also have a vested commercial interest in maintaining these separate knowledge worlds so, not surprisingly, second waves of knowledge building, and abductive reasoning from crossing the boundaries of different knowledge worlds take time. This second wave usually seeks to develop more systematic insights about issues and concepts, it responds to ambiguities and conceptual flaws, and identifies practical boundaries that have emerged from observing real-world implications. Theory testing, applications in different scenarios, and diverse contexts permit deeper insights and a richer understanding of key values and concepts.

Table 2. Key terminologies in 'co-creation'.

Terminologies	Potential contribution in understanding tourism	References
Service studies		
Service-dominant logic Experience co-creation Collaborative transactions	• Reframes **consumers as active participants** in the production of the tourism experience • Moves the concept of **value beyond money** to include other kinds of social, psychological value	Ballantyne and Varey (2006), Blaschke, Haki, Aier, and Winter (2018), Etgar (2008), Mauss (1970), Slee (2015), Vargo and Lusch (2004)
Planning		
Collaborative planning	• **Inclusive and democratic participation** in the planning of tourism • Attributes the reflexive **planner with responsibility** for collaborative, co-creative dialogue	Jamal and Getz (1995)
Communicative action	• Nurturing discursive practices that enhance understanding and identification of **collaborative actions**	Fischer (2012), Healey (1997), Innes and Booher (2007)
Development studies		
Community empowerment Bottom of Pyramid (BOP)	• Reframes the poor as active, involved, informed consumers to co-create the market around the needs of the poor.	Prahalad, 2005
Post-human perspectives		
Hybrid actors, human/ non-human actors	• **De-centres the human perspective** • **New hybrid actors** produced through complex physical, economic, environmental, and human interactions	Akrich, Callon, and Latour (2002), Haraway (2013)
Business Systems/Organisational studies		
Multi-sided platforms	• New business models or strategies that coordinate the demands of two or more groups of producers/customers who have a **mutual/ shared interest in collaboration**	Evans (2003), Olson, Lee, and Trimi (2012)
Prosumer/Prosumption	• Person who produces also consumes, co-creating the product. Emphasis on the **breakdown in traditional models** of supply /demand and value creation at the source of production; defined as the integration of physical activities, mental effort and socio-psychological experiences.	Toffler (1980), Tapscott and Williams (2006), Rifkin (2014), Ritzer and Jurgenson (2010)
Open design	• **Collaborative labouring** on projects of **shared value**, often motivated by need to **spread/reduce cost of resources** (time, expertise, financial support, etc.)	Benkler and Nissenbaum (2006)
Crowdsourcing	• **Collaborative sourcing of information, ideas, money** or other kinds of input by enlisting the services of a crowd of people, either paid or unpaid, typically via the Internet	Estellés-Arolas and González-Ladrón-De-Guevara (2012)
Crowdfunding	• **Collaborative funding** of projects where donors have instrumental need or emotional connection to contribute voluntarily to the cause.	Quero, Ventura, and Kelleher (2017)
Commons economy	• Open systems where contributors **co-create value which contributes to the commons** (i.e. social, cultural and environmental resources held in common and not privately owned)	Dredge and Gyimothy (2017)
Innovation		
Open innovation	• Leveraging of both **internal organizational resources and external resources** as part of the innovation process. Emphasis is on bridging, sharing resources across **organisational boundaries.**	Chesbrough, 2003
Open source	• **Mass collaborative production**, and new technologies such as blockchain, transforming notions of **ownership** of what is produced (e.g. IP), often kicked-started by the broader community	Lerner and Tiroli (2001), Mauss (1970)
Systems of innovation	• **Interactions among actors and networks** contribute to the creation of new ideas and innovation e.g new product, service, process	Lundval (1985), Ind and Coates (2013)
Governance		
Creating shared value (CSV) Corporate social responsibility (CSR) Blended value Public-private governance	• Collaboration between public-private actors can produce outcomes of **mutual or shared value** • Blended value acknowledges a **full register of social, economic, environmental, political value** that can be produced • Emphasis is predominantly on the **motivations to collaborate and mutual interest outcomes.**	Crane et al. (2014), Emerson (2003), Dodds and Joppe (2005)

An alternative perspective on knowledge creation is offered by Foucault (1969, 1970, 1980), who asserts messy, post-structural archaeology of knowledge, where knowledge is socially constructed through multiple, overlapping, sometimes contradictory discourses. There are unities and discontinuities in knowledge formation, different scales at which knowledge coalesces, and crises and/or dominant values, such as neoliberalism, serve to empower some ideas over others (Dredge & Jamal, 2015). This messy context in which scientific knowledge is developed is important in examining the evolution of tourism co-creation.

As a new and fashionable term, co-creation has emerged as a heuristic metaphor, or a cogent schema, that helps to (1) capture in broad elements to explain a phenomenon, and (2) to project values about what is important to the knowledge community. Kuhn (1970) further argues that while the values embedded in a particular paradigm might be shared in the broadest sense, interpretation and application of these values might vary across knowledge domains due to different interests held by the researchers undertaking those reviews. The above systematic reviews of co-creation (Table 1) all take an instrumental approach, starting with the key terms, executing database searches and analysing themes. These reviews, while recognising the diverse threads and themes that exist in the body of works they analyse, do not acknowledge historical roots, their own disciplinary biases, or philosophical stance. While some offer caveats that limit the extent of their data harvesting, or describe other methodological limitations, none of these reviews acknowledge disciplinary biases or limitations. Following on from the systematic analyses above, and critical questioning of the silences and hidden perspectives, a mapping of disciplinary contributions shown in Figure 1 was produced.

Co-creation – a business management perspective

de Oliveira and Cortimiglia (2017) define co-creation as the 'joint, collaborative, concurrent and peer-like practices aimed at creating new types of value'. Prahalad and Ramaswamy (2000, 2004) observe that growth and value creation are two key themes preoccupying most business managers, explaining co-creation as 'the joint creation of value by the company and the customer; allowing the customer to co-construct the service experience to suit their context' (Prahalad and Ramaswamy 2004, p. 8). These authors argue that the meaning of value and mechanics of value creation were shifting from, firstly, a narrow monetary definition of value to include other diverse kinds of value. For example, a visitor experience that fulfils a lifetime ambition produces psycho-social value which is difficult, even impossible, to measure in dollar terms. Secondly, the point at

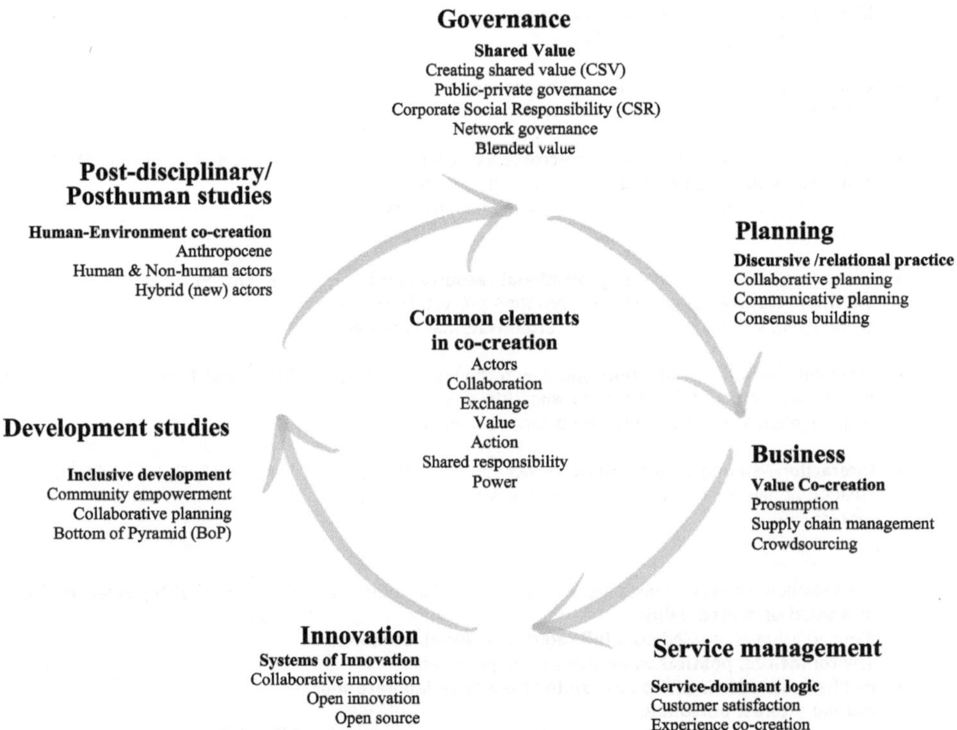

Figure 1. A disciplinary mapping of co-creation.

which value is being created in the value chain is shifting from the traditional view that value is created solely by producers who then need to convince consumers of its value to them. Instead, it is increasingly accepted that the exchange process is more complex and that customers are also producing something of value (such as reviews, testimonials, and images of their experience) within the transaction process.

In acknowledging this trend, the term 'prosumer' – a person who is simultaneously a consumer and a producer was first coined by American futurist Toffler (1980). Its related term 'prosumption' or 'production by consumers' was later made popular during the dot-com era in the 1990s (Ritzer & Jurgenson, 2010). Along a similar line of thoughts, Cova and Dalli (2009, p. 333) proposed the term 'working consumers' to indicate 'the phenomenon of consumers who, by the means of immaterial labour, add cultural and affective elements to market offerings'.

Facilitating this process of value co-creation are advances in information technology, where for example, the rise of social media (e.g. Facebook, Twitter, Instagram, etc.) helps destination and experience marketers build brand awareness, market trust, and through feedback mechanisms, contributes to more responsive and agile product development (de Oliveira & Cortimiglia, 2017). Similarly, the rise of the Internet of Things and online communities have also stimulated the rise of 'crowdsourcing', typically the sourcing of information, money or other kinds of input from a large crowd of people (Estellés-Arolas & González-Ladrón-De-Guevara, 2012). In tourism, this trend is reflected in the increase of online travel information brokers such as TripAdvisor and Wikitravel.

Co-creation – a service-centred perspective

This observation, that value is created at various points in the exchange process, was framed and justified as the key to achieving traditional business values including market expansion, growth, profit maximisation, and supply chain innovation. As a consequence, business managers and marketers have taken an interest in the nature of exchange, and seek to identify opportunities for new value creation. Building upon Prahalad and Ramaswamy's ideas (2004), various scholars such as Vargo and Lusch (2004, 2006, 2008a, 2008b), Frow and Payne (2007), Payne, Storbacka, and Frow (2008), Cova and Dalli (2009), Cova, Dalli, and Zwick (2011) and Schmitt (2010) have contributed to develop two emergent streams of work – service-dominant (S–D) logic and service science – in an effort to co-create a more marketing-grounded understanding of value and the characteristics of exchange that goes beyond the traditional goods-dominant (G–D) logic.

In tourism studies, and based on S–D logic, Prebensen, Kim, and Uysal (2016, p. 1) define co-creation of value 'as the tourist's interest in mental and physical participation in an activity and its role in tourist experiences'. This application of S–D logic in tourism has led to a large and growing body of work that has sought to explore the role of the customer as an actor in the creation of tourism experiences (Campos et al., 2016). Research has tended to reinforce findings that active participation of tourists in the visitor experience enhances visitor satisfaction (e.g. Buonincontri et al., 2017; Prebensen et al., 2016). This view is largely underpinned by the theoretical framework of the experience economy, which posits that a focus on creating personal value for consumers (e.g. personal branding, social connections, transformative changes in physical or mental beliefs of tourists) will subsequently lead to an increase in economic value for marketers and providers (e.g. Pine & Gilmore, 1998). Subsequently, Customer Experience Management – a comprehensive set of frameworks, tools, and methodologies that enables firms to co-create and manage customer experiences – has increasingly been researched and applied in tourism and hospitality context (see e.g. Kandampully, Zhang, & Jaakkola, 2018)

Beyond a customer focus, the service science literature suggests a more holistic view of co-creation as being embedded within the service systems. This view focuses on the diverse combinations of resources (often consisting of human capital (people), ICT (technology) and networks of firms) that enable the co-creation of value to take place (Saarijärvi et al., 2013). In the business and service management literature, 'collaborative transaction' emerges as an umbrella term that encompasses various hybrid market models and conceptualisation of resources such as Peer-to-Peer (P2P), Business-to-Business (B2B), Customer-to-Business (C2B), Customer-to-Customer (C2C) and many-to-many marketing (Saarijärvi et al., 2013).

Co-creation – an innovation-centred perspective

On a similar note to the service system view, the 'systems of innovation' framework developed by B. A. Lundvall (1985) emphasises that new ideas emerge between, rather than within people and that co-creation practices between actors within a system is the key to ensuring that a system remains innovative and competitive. Consequently, increased attention has been placed on identifying and unlocking new value creation opportunities, and in catalysing new products and experiences based on collaboratively rethinking business ecosystems, distribution channels, markets, and so on. This has gone hand-in-hand with increased policy emphasis on digitalisation and e-tourism (Cabiddu et al., 2013).

Tekic and Willoughby (2018, p. 15) conducted an extensive systematic review of the innovation literature and defined co-creation as 'a form of *collaborative innovation* initiated by a company, involving individual external contributors or co-creators – not just users and customers, but also field experts, students, or amateur innovation enthusiasts – who may provide valuable input to the company's innovation projects'. This definition represents a paradigm shift in innovation practices, partly influenced by 'open innovation' and 'open source' movements within the information technology community in the 1990s (Coughlan, 2013; Ritzer & Jurgenson, 2010). More recently, the term 'open innovation' has been adopted by the business community and made popular by Chesbrough (2003), whose work focused on unleashing the competitive advantage of individual firms through open innovation. In contrast to 'closed innovation' models, where firms innovate primarily through internal research and development (R&D), open innovation emphasises the leveraging of both internal and external resources (e.g. knowledge, technology, people) as part of an innovation process. Similarly, the 'open source' movement operates on the premise of reciprocal exchange, where the mass co-creation process is often kick-started with a 'gift' or a generous offering to the broader community (Mauss, 1970). Other closely related terms are 'open design', and 'common-based peer production', where, due to a lack of commercial interest or funding, people invest skills, time or other inputs into projects for the common good (Benkler & Nissenbaum, 2006).

In tourism, this innovation co-creation lens has enabled managers to unlock additional and diverse forms of value through interactions between hosts, customers, digital platforms, DMOs, and businesses (de Oliveira & Cortimiglia, 2017). For instance, rapid growth of *multi-sided collaborative platforms* such as home, restaurant, and ride-sharing platforms, illustrate the innovative and disruptive impact of these developments, which are now transforming the tourism sector (Belk, 2010).

Co-creation – a governance perspective

Just as co-creation was taking off as an exciting development in business management and service studies, in 2008 the Global Financial Crisis raised questions about the continued dominance of the profit and growth mindset. Critical questions started to re-emerge over hyper-capitalism, corporate greed, the corporate sector's lack of moral code, and increasing inequity between the rich one percent and the remainder of the world's population. Furthermore, these issues were coupled with increasing concerns over climate change, loss of biodiversity, and other environmental problems. This discord created the context for a wave of disruptive thinking, and a flurry of research on collaborative governance and the concept of *Creating Shared Value* (CSV).

Porter and Kramer (2011) have been given credit for popularising the term CSV, which captures the idea that, in order for business to regain trust and legitimacy, they need to pursue values that are shared with society at large. They argue for new ways of framing and pursuing business growth by identifying ways in which societal issues can be addressed while simultaneously pursuing traditional profit-making activities (Crane, Palazzo, Spence, & Matten, 2014). They proposed three main strategies for creating shared value: by re-conceiving products and markets; by redefining productivity in the value chain; and enabling cluster development. In many ways, Porter and Kramer were repackaging old ideas, reiterating ideas about more ethical and sustainable forms of capitalism, reasserting school of thoughts such as corporate philanthropy and Corporate Social Responsibility (CSR), and drawing inspiration from emergent concepts such as impact investing, and blended value (Dodds & Joppe, 2005; Emerson, 2003). Furthermore, the concept of shared value is not unique to the domain of business governance but draws its root and inspiration from a range of established research areas such as of public governance, participatory governance, the pursuit of mutual benefit (e.g. Fischer, 2012)

Despite its increased popularity in the literature, there have been strong criticisms that CSV is capitalism as usual, or that it even gives license to a new and more pervasive phase of capitalism where social and environmental problems are folded into, and silenced under, capitalism's pursuit of growth and profit (Crane et al., 2014). Within the area of public governance, the increased emphasis on privatisation of public assets/services, and for public decisions and actions to be made in alignment with private sector interests, have been criticised as supporting neoliberal ideas about economic value creation, whilst neglecting alterative value that are important for individual and community wellbeing but not necessarily recognised by market logics (e.g. community cohesion, ethics of care) (Bauwens, 2006). In tourism, this may manifest in the co-creation of tourism policies that reinforce the sector's growth agenda instead of challenging the status quo to reduce inequality and other social-environmental issues related to tourism development in local communities.

Co-creation – a planning and policy-making perspective

Although terminology differs from the business and management literature, the intention of building shared collaborative approaches through communicative and consensus building approaches is reminiscent of the more recent 'co-creation' terminology. Collaborative planning emphasises the importance of moving beyond tokenistic consultation with local actors to empower such communities as active agents. Early antecedents can be found in the activist work of Jane Jacobs (1961), the advocacy work of Davidoff (1965), the collaborative planning approach developed by Healey (1981, 1997), communicative relational approaches to policy advocated by Fischer (2012), and the consensus building work by Innes and Booher (2007). Collaborative planning emphasises a shift in the dominant approach to planning, from a rational scientific to a relational approach. The work of these diverse authors frame collaborative planning as inclusive, interactive, democratic, communicative, pluri-vocal, and action-oriented (Healey, 1998), all values that resonate with contemporary ideas of co-creation.

In tourism, these influences flowed through to Jamal and Getz's (1995) seminal work on collaborative tourism, and discussions of networked, collaborative, and participatory governance (e.g. Araujo & Bramwell, 1999; Bramwell & Lane, 2000; Reed, 1997). Participatory/collaborative governance emphasises the deepening of citizens' democratic engagement in the governmental processes, empowering them to undertake various roles (e.g. as co-implementer, co-designer, and initiator of actions) which places them at the centre of grass-root social innovation (Voorberg, Bekkers, & Tummers, 2013). Most of this work adopts a place-based community approach to co-creating local actions, which is understandable due to the disciplinary links between planning and geography. What is important, however, is the long and well-established links with communicative action, drawing upon Habermas, and Bourdieu's concept of habitus and communities of practice, that has the potential to inform current discussions of co-creation.

Co-creation – a development perspective

In development studies, growing criticism of non-profit and public sector ineffectiveness in addressing many persistent developmental issues (e.g. poverty, marginalisation) has led to the criticism that top down approaches to international aid are ineffective, and that inclusive, bottom up community driven initiatives are needed (Sharpley, 2009). Advocates of the Bottom of Pyramid (BOP) approach call for context-based solutions to be co-created with people at the bottom of the economic pyramid. Adapting the logics of traditional business management, Prahalad (2005, p. xii) believes that by reframing the billions of people who earn less than two dollars a day as 'active, informed and involved customers', lasting positive changes will result from 'co-creating the market around the needs of the poor'.

One of the most frequently cited examples of this BOP approach is microfinance. Pioneered by Professor Mohamed Yunus and the Grameen Bank of Bangladesh in the 1970s, microfinance is the provision of small-scale financial services such as micro-credit, micro-saving and micro-insurance to address the financial needs of people living in poverty, who would usually be excluded from the formal financial institutions (Schreiner & Colombet, 2001). Through charging sufficient interest rates to cover the operating costs, the global microfinance industry (now worth over 100 billion US dollar) demonstrates that poor people are indeed a very important consumer market (Helms, 2006).

In the context of tourism, the last few decades have witnessed the rapid rise of globalisation and a 'new mobilities paradigm', yet there remain billions of people living in poverty who are socially, politically and economically excluded from travel (Hall, 2010). For this so-called 'immobile' population, the ability to travel for any period of time and for any length of distance (even just from their home village to the nearest city) means much more than a leisure experience. It may open up, among other things, access to proper medical and legal services, education and economic opportunities and new livelihood ideas that are instrumental for a better life. Rogerson (2014) for instance, called for more attention and support to the 'migrant entrepreneurs' or the 'necessity tourists' who travel almost daily across borders in sub-Saharan Africa in order to make a living through subsistent trading.

Arguably, the BOP approaches to co-creating inclusive economic opportunities and affordable travel products and services have potential to deliver significantly more value to people at 'the bottom of the pyramid'. The BOP approach is also supported by the advances of digital technologies, which have fuelled the rise of informal and sharing economy. These may include online sharing platforms that offer more affordable ride-sharing, ride-hailing and accommodation services to people with lower incomes (Dredge & Gyimothy, 2017). Besides platform capitalism (e.g. Airbnb, Uber), tourism non-profit cooperative platforms such as Fairbnb, Authenticitys, and VolunteerMatch have contributed to enable the local citizens with significantly less resources to participate in and gain benefits from the tourism system, in turn creating and distributing value in fairer ways. However such platforms have struggled to establish viable business models to date (see e.g. Bauwens, 2006; Scholz, 2016).

Co-creation – a post-human perspective

Science and technology studies have pushed the boundaries of co-creation in another direction, arguing that, firstly, non-human elements such as ideas, things, artefacts and so on can also have agency (Callon, 1998; Latour, 2005). In tourism contexts, for example, natural environments and urban landscapes provide a backdrop to the perfect Instagram photo, helping to co-create a visitor's identity, contributing to their visitor experience, but also activating the audience to take action. These human and non-human actors co-produce visitor satisfaction through a joint collaborative process that takes place between human and non-human actors.

Second, joint, collaborative and co-creative processes between people, objects, ideas, and other things can create hybrid actors, or what Haraway describes as technology-infused humans or cyborgs (2013). These ideas are challenging for some, and it is not the role of this paper to offer any detailed critique. However, these diverse contributions, including the work of Star and Strauss (1999) and Star (2010) on boundary objects, Callon (e.g. 1998, 2006) on the performativity of economics, and Donna Haraway's (2013) description of shifting coalitions of more-than-human actors suggest that the traditional categories of things are fusing, coalescing, hybridising, and taking on new meaning. Traditional ways of understanding agency as human-centred, and categorising and organising our understanding using a very human perspective are being challenged. So, in the context of mass tourism, it is possible to interpret the notion of swarming crowds as an emergent, but more-than-human actor in its own right. The crowd is an assemblage of people, of images, of sensory cues, of visual and textual artefacts, and of psycho-social reactions and responses that is more than the sum of its human parts. Technology interacts with human visitors shifting and shaping the way the crowd behaves and responds. But there are also invisible and silent components such as environmental damage and declining ecological health caused by mass travel. Together these elements contort and transform the pulsating crowd into something that is both more-than-human and interscalar, with visible and invisible parts. Callon (2006) highlights that the discourses around such actors, in this case, a swarming crowd, are performative – they produce what they describe. The language of overtourism defines how the crowd is viewed, perceived, understood, and the largely negative value that is created.

Recent thinking by those engaging with the challenges of the Anthropocene and the Rights for Nature movement also suggest that we need to acknowledge the co-created value generated through the interaction between non-human and human actors (Lund, 2013). In other words, sustainability in tourism relies on more inclusive and holistic approaches to value co-creation that extend well beyond the 'business-as-usual' mindset, and that takes into account the contributions of non-human actors as diverse as nature, silence, carbon, and so on.

Key concepts and elements of co-creation

In the above text, we have traced seven threads of scholarship that demonstrate the ideas and values associated with co-creation have diverse historical roots in a range of disciplines and fields of study. There may be more disciplinary threads that we have not identified, or that we are not aware of due to our own knowledge limitations. However, it is clear even among the threads that we have traced, that they employ overlapping ideas, similar concepts, and are interdependent like the image of a DNA sequence that we invoked earlier. For instance, while innovation-centred perspective is represented as a separate theme, both business management and service-centred perspectives have frequently taken innovation into account in their interpretation of co-creation. The work of advancing understanding co-creation and its relevance to tourism, and our aim of excavating novel and fresh approaches, lies in this interdisciplinary enterprise (e.g. see Stember, 1991).

The discussions above illustrate that concept of co-creation draws from the relational turn in sociology, and emphasises a number of common elements including that it is collaborative, communicative, discursive, relational, action-oriented, participatory, democratic, inclusive, and so on. Of course, different disciplinary leanings mean that the different values and motivations of co-creation are highlighted, and the contributions of different ontologies and diverse methodologies are recognised. A synthesis of these different approaches assists in a more complete understanding of the whole. Our departure point from the start was that co-creation, as a fashionable term and metaphor, is an abstraction open to interpretation, and stops short of informing scholars how they might work together. An interdisciplinary meta-narrative analysis helps to identify the overlapping ideas, it transcends boundaries to identify similar concepts and terms, and it helps to mesh together diverse thinking and helps scholars take back certain ideas and elements back into their own thinking and disciplinary communities.

Stember (1991) suggests that the interdisciplinary enterprise asks that we integrated knowledge and methods from different disciplines and fields of study, acknowledging distinctions and contributions. She identifies three potential lines of enterprise: developing the intellectual domain, exploring the practical implications; and implications for the pedagogical domain. This is a much larger project, and much of it is outside our immediate aims in this paper. However, in what follows, we identify, transcend boundaries, and extend the key concepts and logics of co-creation and make a contribution to a fresh and novel approach to co-creation that cuts across disciplinary divides and points attention towards a more holistic approach to understanding the co-creation of value in tourism.

Discussion

Our aim in this paper was two-fold: Firstly, we sought to critically examine and extend the notion of collaborative tourism making by undertaking a meta-narrative analysis of co-creation. Secondly, we sought to identify and extend key concepts of co-creation, and in the process build and understanding of co-creation as something broadly relevant to all tourism researchers, not just those working with service-dominant logic, visitor experience and marketing. In addressing these two aims, the intention was not to apply a homogenising filter over the diverse thinking that has gone into co-creation and related concepts, but rather, to acknowledge overlapping and interwoven historical roots, related terms, and thinking. We chose not to define the term at the beginning of the paper, but rather, let the meta-narrative unfold and draw it together here in the discussion. Etymology provides the key to understanding, where co-creation is the act of creating something together. The meta-narrative also provides insights into how co-creation is discussed and interpreted in the diverse literature (Table 2 summarises this diversity), and from this, it is clear that different ontological, epistemological, methodological and axiological influences mean that there can be no consensus on an overarching definition for tourism studies. That said, however, we can draw attention to seven key features, which also point to a rich research agenda for the future:

(1) *Co-creation involves value creation.* Value is a complex concept. Creating value – money, resources, labour, shareholder value and so on – is a traditional objective of neoclassical economics. However, discussions of co-creation highlight that other forms of social, cultural, political and environmental value can also be produced, and that these are balanced against financial gain when consumers make decisions. Value is also dynamic, slippery, fleeting or permanent, and can be conceived of as an object, an aim, an outcome, or a process. We need to better understand it, from the perspective of visitors, residents, destinations, organisations, and non-human actors like nature, and so on.

(2) *Co-creation involves two or more actors or actor groups producing something together.* The roles and responsibilities of various actors involved in co-creation in tourism are challenging prevailing ideas about the tourism system, and traditional roles and responsibilities, e.g. consumers and producers. We need to better understand co-creation contributes to new understandings of the tourism ecosystem.

(3) *Co-creation involves the collaborative exchange of resources such as time, energy, money, expertise, and so on.* Digital technologies are mooted as a way of facilitating these exchanges, but co-creation is more than technology. The broader influences of the techno-anthropological landscape of co-creation need to be better understood.

(4) *Co-creation unleashes new models of collaboration, sharing, gifting, access, and other kinds of transactions often sidelined in neoclassical economics.* While much celebrated in the literature, there is a dark side to co-creation. Not all co-creation activities are consensual, and non-consenting parties (such as residents in a neighbourhood overtaken by Airbnb) may be excluded or their interests are not considered. Market failures associated with co-creation need to be better understood.

(5) *Co-creation is political.* The very act of collaboratively creating something is a political act where actors exercise their agency. The planning and governance literature, in particular, drew attention to the inclusive, democratic, outcomes associated with co-creation. But, in the point above, sometimes parties experiencing the impacts are excluded. We need to better understand the 'who wins' and 'who loses' in co-creation.

(6) *Co-creation has given rise to new/hybrid actors.* The de-centring of humans in processes of collaborative co-creation, has shone a light on how actors can be thought of in fresh ways, helping to rethink traditional approaches and reconceptualize key challenges. We need to incorporate the rights of non-human actors, such as nature, forests, rivers and so on, where such innovative thinking can disrupt

traditional thinking and help reformulate the challenges we face.

(7) *Co-creation is closely associated with contemporary ideas about innovation*. Innovation in systems of production and consumption, in business ecosystems and supply chains, in processes and practices, have emerged as a result of collaborative ways of working together. Co-creation (sharing, collaboration, gifting, etc.) has redefined how we access resources such as knowledge expertise, capital, labour, and so on. Economic geographers have highlighted that the opposite of co-creation can be lock-in, where innovation is hampered by inability to share collaborate, remove institutional impediments. We need to understand more about how co-creation may enhance innovation through inclusive thinking, or impeded it through exclusive (invitation only) co-creation practices.

(8) *Co-creation is transforming ideas about who/what owns the value produced, and who has responsibility for its management/stewardship*. The collaborative co-production of something of value may come about as a result of resource pooling, sharing and contributing freely to a common goal. Co-creation raises questions about the collaborative commons and the management of resources that are owned by no one in particular. We need to know more about the potential of the collaborative commons, how to manage it and in whose interests.

(9) *The relational characteristics of actors involved have an important impact on the co-creation process and outcomes* (e.g. ethics, motivations, emotions, power, equality). Unbalanced and unstable power relations due to privilege, information or resource asymmetries can potentially lead to value co-destruction instead of co-creation (Echeverri & Skålén, 2011). The 'how' of co-creation process, therefore, cannot be separated from the awareness and acknowledgement of the 'what' (i.e. what kind of value is created) and the 'who' (i.e. who participates and who benefits from the created value). We need to better understand the relational work involved in co-creation.

Conclusions and future research

There is no doubt that co-creation is a fashionable concept. In tourism, co-creation has predominantly been examined and theorised within a business and service context, and its typically human-centred, and focused on value creation that sustains and promotes existing capitalist forms of economic activity. The meta-narrative revealed that co-creation has been reduced to a utilitarian value-producing concept between categories of actors, (e.g. producers and consumers or hosts and guests). Arguably, this narrow conceptualisation of co-creation in tourism can marginalise broader discussions of collaborative and co-created actions that exist across different strands of literature. Tourism is much more than economic value-producing transactions but can also generate alternative kinds of value, both positive and negative, that influence local wellbeing, liveability and flourishing, place attachment, resource protection and conservation, confidence in the future, migration, international relations and macro-economic management. It is precisely this complexity that makes the concept of co-creation an interesting and useful lens for building a multidisciplinary understanding of tourism and how it changes people, places and things.

We recognise in this meta-narrative analysis that co-creation also has deeper roots in notions of civil society and democracy. Our approach to and interpretation of co-creation either provides or hinders access to the structures and processes through which we are governed. Accordingly, co-creation can also be understood as a much older dimension of the cooperative and collaborative human condition. Co-creation, collaboration, shared production, partnering, and co-operation similarly capture the idea that value is produced by working together. It was our ambition in this paper to transcend the ontologies that have created and fed these different streams of research, and to recognise that working together to produce diverse kinds of value, understanding, and collaborative outputs and/or actions for diverse actors is fundamental to addressing the range of challenges that we face. For example, hosts work with visitors, communities work with industry, producers work with consumers, governments work with industry, NGOs work with volunteers, and industry must work with environmental actors, to co-produce diverse outcomes which might be valued in vastly different ways. In doing so, categories of things become blurred and dynamic, where, for example, community members become experts, researchers become learners, problems become opportunities, for example.

In using a meta-narrative approach to unpack co-creation, we have shifted traditional conceptualisations of tourism co-creation by (1) expanding the concept to include diverse forms of social, political, cultural, and environmental value; (2) expanding ideas about who produces and who benefits from that value; (3) exploring the resources that are used or consumed in the creation of that value; and (4) raised questions about who wins and who loses in value creation. The use of a broader interdisciplinary framework of value co-creation provides an analytical lens that directs attention to issues of participation, inclusion and distribution of costs and benefits of tourism, which contribute to new and fresh appreciation of value co-creation.

These above questions hopefully can provide the basis for a more comprehensive examination of value co-creation in future tourism research. In this way, co-creation can help to unleash tourism's potential as a powerful co-creative social force, as opposed to extractive industry. As Ind and Coates (2013, p. 92) argue, 'co-creation can be a force for participation and democratisation that does create meaning for all, rather than simply an alternative research technique or a way of creating value through co-opting the skills and creativity of individuals'.

In sum, we leave readers with a nine-point research agenda drawn from the above identified characteristics of co-creation and our interdisciplinary meta-narrative review. For those wishing to adopt co-creation practices, the following points of consideration should guide bespoke co-creation future research approaches and implementation:

(1) Future approaches should consider co-creation from multiple perspectives (also known as personas, avatars, etc.) including visitors, residents, destinations, organisations, and non-human actors like nature, animals, and so on.
(2) Future co-creation approaches should consider how new and fresh understandings of the tourism ecosystem, including human and non-human components can be unearthed.
(3) Future co-creation approaches to problem solving should consider broader influences of the techno-anthropological landscape, and the power of technology in co-creating tourism.
(4) Market failures or any negative impacts associated with co-creation should be considered and steps taken to minimise in the process.
(5) We should consider who wins and who loses in co-creation practices and define inclusive co-creation principles relevant to the context in which we use co-creation approaches.
(6) We should consider co-creation from the perspective of non-human actors, such as nature, forests, rivers and so on, where such innovative thinking can disrupt traditional thinking, help reformulate the challenges we face, and manage co-creation so that the interests of those without a human voice are also protected.
(7) We should consider how co-creation may enhance innovation by being inclusive, or impeded it through exclusive (invitation only) co-creation practices.
(8) We should consider the potential of the collaborative commons, how to manage the commons, and in whose interests should it be managed.
(9) We should consider the relational work involved in co-creation, and the costs and benefits of co-creation for different (human and non-human) actors.

Together, these points contribute to the future research agenda and implementation of co-creation practices in tourism. So whether it is a local tourism organisation, a business, or a community group seeking to address a tourism related challenge in a collaborative manner, the points above prompt us to carefully consider, anticipate, and articulate how co-creation might be used as an effective and inclusive approach to joint action.

Disclosure statement

No potential conflict of interest was reported by the authors.

ORCID

Giang T. Phi http://orcid.org/0000-0002-2359-2833

References

Akrich, M., Callon, M., & Latour, B. (2002). The key to success in innovation part 1: The art of interessement. *International Journal of Innovation Management*, *6*, 187–206.
Alves, H., Fernandes, C., & Raposo, M. (2016). Value co-creation: Concept and contexts of application and study. *Journal of Business Research*, *69*(5), 1626–1633.
Araujo, L. M., & Bramwell, B. (1999). Stakeholder assessment and collaborative tourism planning: The case of Brazil's Costa Dourada project. *Journal of Sustainable Tourism*, *7*(3–4), 356–378.

Ballantyne, D., & Varey, R. J. (2006). Creating value-in-use through marketing interaction: The exchange logic of relating, communicating and knowing. *Marketing Theory, 6*(3), 335–348.

Bauwens, M. (2006). The political economy of peer production. *CTheory*. Retrieved from https://journals.uvic.ca/index.php/ctheory/article/view/14464/5306

Belk, R. (2010). Sharing. *Journal of Consumer Research, 36*(5), 715–734. doi:10.1086/612649.

Benkler, Y., & Nissenbaum, H. (2006). Commons-based peer production and virtue. *Journal of Political Philosophy, 14*(14), 394–419.

Blaschke, M., Haki, M. K., Aier, S., & Winter, R. (2018). *Value co-creation ontology – a service-dominant logic perspective*, Multikonferenz Wirtschaftsinformatik 2018, March 06–09, 2018, Lüneburg, Germany.

Bramwell, B., & Lane, B. (2000). *Tourism collaboration and partnerships: Politics, practices and sustainability*. Clevedon: Channel View.

Buonincontri, P., Morvillo, A., Okumus, F., & van Niekerk, M. (2017). Managing the experience co-creation process in tourism destinations: Empirical findings from Naples. *Tourism Management, 62*, 264–277.

Cabiddu, F., Lui, T.-W., & Piccol, G. (2013). Managing value co-creation in the tourism industry. *Annals of Tourism Research, 42*(1), 86–107.

Callon, M. (1998). *The laws of the markets*. London: Blackwell.

Callon, M. (2006). *What does it mean to say that economic is performative*? CSI Working Papers.

Campos, A. C., Mendes, J., do Valle, P. O., & Scott, N. (2016). Co-creation experiences: Attention and memorability. *Journal of Travel & Tourism Marketing, 33*(9), 1309–1336.

Chesbrough, H. (2003). The era of open innovation, *MIT Sloan Management Review*. Spring, 35–36.

Čorak, S., & Živoder, S. (2017). Tourism destination and DMO transformation. In L. Dwyer, R. Tomljenović & S. Čorak (Eds.), *Evolution of destination planning and strategy* (pp. 99–118). London: Palgrave Macmillan.

Coughlan, S. (2013). *Thoughts on open innovation essays on open innovation from leading thinkers in the field*. Brussels: OpenForum Europe.

Cova, B., & Dalli, D. (2009). Working consumers: The next step in marketing theory? *Marketing Theory, 9*(3), 315–339.

Cova, B., Dalli, D., & Zwick, D. (2011). Critical perspectives on consumers' role as 'producers': Broadening the debate on value co-creation in marketing processes. *Marketing Theory, 11*(3), 231–241.

Crane, A., Palazzo, G., Spence, L., & Matten, D. (2014). Contesting the value of 'creating shared value'. *California Management Review, 56*(2), 30–49.

Davidoff, P. (1965). Advocacy and pluralism in planning. *Journal of the American Institute of Planners, 31*, 331–338.

Dodds, R., & Joppe, M. (2005). *CSR in the tourism industry? The status of and potential for certification, codes of conduct and guidelines*. Washington: IFC/World Bank.

Dredge, D., & Gyimothy, S. (2017). *The collaborative economy and tourism*. Cham: Springer.

Dredge, D., & Jamal, T. (2015). Progress in tourism planning and policy: A post-structural perspective on knowledge production. *Tourism Management, 51*, 285–297.

Eacott, S. (2018). The relational turn in social sciences. In S. Eacott (Aut.), *Beyond leadership: A relational approach to organizational theory in education* (pp. 25–41). Singapore: Springer.

Echeverri, P., & Skålén, P. (2011). Co-creation and co-destruction: A practice-theory based study of interactive value formation. *Marketing Theory, 11*(3), 351–373.

Emerson, J. (2003). The blended value proposition: Integrating social and financial returns. *California Management Review, 45*(4), 35–51.

Estellés-Arolas, E., & González-Ladrón-De-Guevara, F. (2012). Towards an integrated crowdsourcing definition. *Journal of Information Science, 38*(2), 189–200.

Etgar, M. (2008). A descriptive model of the customer co-production process. *Journal of the Academy of Marketing Science, 36*(1), 97–108. doi:10.1007/s11747-007-0061-1

Evans, D. S. (2003). Some empirical aspects of multi-sided platform industries. *Review of Network Economies, 2*(3). Retrieved from http://www.rnejournal.com/articles/evans_final_sept03.pdf

Fischer, F. (2012). Participatory governance: From theory to practice. In D. Levi-Faur (ed.), *The Oxford handbook of governance* (pp. 457–471). Oxford: Oxford University Press. doi:10.1093/oxfordhb/9780199560530.013.0032

Fleury-Vilatte, B., & Philippe Hert, P. (2003). The boundaries between disciplines. Frontières disciplinaires, 3. Retrieved from https://journals.openedition.org/questionsdecommunication/8855?lang=en

Foucault, M. (1969). *The archaeology of knowledge*. Abingdon: Routledge.

Foucault, M. (1970). *The order of things: An archaeology of the human sciences*. New York, NY: Random House.

Foucault, M. (1980). In C. Gordon (Ed.), *Power/knowledge: Selected interviews and other writings*. New York, NY: Harvester Wheatsheaf.

Frow, P., & Payne, A. (2007). Towards the 'perfect' customer experience. *Journal of Brand Management, 15*(2), 89–101.

Galvagno, M., & Dalli, D. (2014). Theory of value co-creation: A systematic literature review. *Managing Service Quality: An International Journal, 24*, 643–683.

Gough, D. (2013). Meta-narrative and realist reviews: Guidance, rules, publication standards and quality appraisal. *BMC Medicine, 2013*(11), 22.

Hall, C. M. (1999). Rethinking collaboration and partnerships: A public policy perspective. *Journal of Sustainable Tourism, 7*(3), 274–289.

Hall, C. M. (2010). Equal access for all? Regulative mechanisms, inequality and tourism mobility. In S. Cole, & N. Morgan (Eds.), *Tourism and inequality: Problems and prospects* (pp. 34–48). London: CABI.

Haraway, D. (2013). *Anthropocene, Capitalocene, Chthulhucene. Donna Haraway in conversation with Martha Kenney*. Retrieved from https://static1.squarespace.com/static/53e3ff14e4b029aa443dd0af/t/55806ee6e4b0fd31006b7ce0/1434480358842/ARTANTHRO_HARAWAY_PROOF.pdf.

Healey, P. (1981). The communicative turn in planning theory and its implications for spatial strategy formation. *Readings in Planning Theory, 23*(2), 217–234.

Healey, P. (1997). *Collaborative planning: Shaping places in fragmented societies*. Vancouver: UBC Press.

Healey, P. (1998). Collaborative planning in a stakeholder society. *Town Planning Review, 69*(1), 1–21.

Helms, B. (2006). *Access for all: Building inclusive financial systems*. Washington, DC: World Bank.

Ind, N., & Coates, N. (2013). The meanings of co-creation. *European Business Review, 25*(1), 86–95.

Innes, J., & Booher, D. (2007). Consensus building as a role playing and bricolage: Towards a theory of collaborative planning. *Journal of Planning Education and Research, 21*(3), 221–236.

Jacobs, J. (1961). *The death and life of great American cities*. New York, NY: Random House.

Jamal, T. B., & Getz, D. (1995). Collaboration theory and community tourism planning. *Annals of Tourism Research, 22*, 186–204.

Jensen, O., & Prebensen, N. (2015). Innovation and value creation in experience-based tourism. *Scandinavian Journal of Hospitality and Tourism, 15*(1), 1–8.

Kandampully, J., Zhang, T., & Jaakkola, E. (2018). Customer experience management in hospitality. *International Journal of Contemporary Hospitality Management, 30*(1), 21–56.

Kuhn, T. (1970). *The structure of scientific revolutions* (2nd ed.). Chicago: University of Chicago Press.

Latour, B. (2005). *Reassembling the social: An introduction to actor-network theory*. Oxford: Oxford University Press.

Lerner, J., & Tirole, J. (2001). The open source movement: Key research questions. *European Economic Review, 45*(4–6), 819–826.

Lipman, M. (1991). *Thinking in education*. Cambridge: Cambridge University Press.

Lipman, M. (2003). *Thinking in education* (2nd ed.). Cambridge: Cambridge University Press.

Lund, K. (2013). Experiencing nature in nature-based tourism. *Tourist Studies, 13*(2), 156–171.

Lundvall, B. A. (1985). *Product innovation and user-producer interaction: The learning economy and the economics of hope*. London: Anthem Press.

Mandrella, M., Zander, S., & Kolbe, L. M. (2016). IT-based value co-creation: A literature review and directions for future research. In Proceedings of *49th Hawaii International Conference on System Sciences (HICSS)* (pp. 287–296). IEEE.

Mauss, M. (1970). *The gift: Forms and functions of exchange in archaic societies*. London: Cohen & West.

Normann, R., & Ramirez, R. (1993). From value chain to value constellation: Designing interactive strategy. *Harvard Business Review, 71*(4), 65–77.

de Oliveira, D. T., & Cortimiglia, M. N. (2017). Value co-creation in web-based multisided platforms: A conceptual framework and implications for business model design. *Business Horizons, 60*(6), 747–758.

Lee, S., Olson, D., & Trimi, S. (2012). Co-innovation: Convergenomics, collaboration, and co-creation for organizational values. *Management Decision, 50*(5), 817–831.

Paul, R. (2011). *Critical thinking movement: 3 waves*. Retrieved from https://www.criticalthinking.org/pages/critical-thinking-movement-3-waves/856

Payne, A. F., Storbacka, K., & Frow, P. (2008). Managing the co-creation of value. *Journal of the Academy of Marketing Science, 36*(1), 83–96.

Phi, G., Dredge, D., & Whitford, M. (2014). Understanding conflicting perspectives in event planning and management using Q method. *Tourism Management, 40*, 406–415.

Pine, B. J., & Gilmore, J. H. (1998). Welcome to the experience economy. *Harvard Business Review, 76*, 97–105.

Porter, M. E., & Kramer, M. R. (2011). Creating shared value. *Harvard Business Review, 89*(1–2), 62–77.

Powell, C., & Dépelteau, F. (Eds.). (2013). *Conceptualizing relational sociology: Ontological and theoretical issues*. New York, NY: Palgrave McMillan.

Prahalad, C. K. (2005). *Fortune at the bottom of the pyramid*. Upper Saddle River: Wharton Business School & Pearson Education.

Prahalad, C. K., & Ramaswamy, V. (2000). Co-opting customer competence. *Harvard Business Review, 78*(1), 79–90.

Prahalad, C. K., & Ramaswamy, V. (2004). Co-creation experiences: The next practice in value creation. *Journal of Interactive Marketing, 18*(3), 5–14.

Prebensen, N. K., Kim, H., & Uysal, M. (2016). Cocreation as moderator between the experience value and satisfaction relationship. *Journal of Travel Research, 55*(7), 934–945.

Quero, M. J., Ventura, R., & Kelleher, C. (2017). Value-in-context in crowdfunding ecosystems: How context frames value co-creation. *Service Business, 11*(2), 405–425.

Reed, M. (1997). Power relations and community-based tourism planning. *Annals of Tourism Research, 24*(3), 566–591.

Rifkin, J. (2014). *The zero marginal cost society: The internet of things, the collaborative commons and the eclipse of capitalism*. New York, NY: Palgrave Macmillan.

Ritzer, G., & Jurgenson, N. (2010). Production, consumption, prosumption. *Journal of Consumer Culture, 10*(1), 13–36.

Rogerson, C. M. (2014). Informal sector business tourism and pro-poor tourism: Africa's migrant entrepreneurs. *Mediterranean Journal of Social Sciences, 5*(16), 153.

Saarijärvi, H., Kannan, P. K., & Kuusela, H. (2013). Value co-creation: Theoretical approaches and practical implications. *European Business Review, 25*(1), 6–19.

Scheurich, J. (1994). Policy archaeology: A new policy studies methodology. *Journal of Education Policy, 9*(4), 297–316.

Schmitt, B. H. (2010). *Customer experience management: A revolutionary approach to connecting with your customers*. Hoboken, NJ: John Wiley & Sons.

Scholz, T. (2016). *Platform cooperativism: Challenging the corporate sharing economy*. New York, NY: Roxa Luxenburg Stiftung.

Schreiner, M., & Colombet, H. H. (2001). From urban to rural: Lessons for microfinance from Argentina. *Development Policy Review, 19*(3), 339–354.

Sharpley, R. (2009). Tourism and development challenges in the least developed countries: The case of the Gambia. *Current Issues in Tourism, 12*(4), 337–358. doi:10.1080/13683500802376240

Slee, T. (2015). *What's yours is mine. New York city*. New York, NY: OR Books.

Snilstveit, B., Oliver, S., & Vojtkova, M. (2012). Narrative approaches to systematic review and synthesis of evidence for international development policy and practice. *Journal of Development Effectiveness, 4*(3), 409–429.

Star, S. L. (2010). This is not a boundary object: Reflections on the origin of a concept. *Science, Technology, & Human Values, 35*(5), 601–617.

Star, S. L., & Strauss, A. (1999). Layers of silence, arenas of voice: The ecology of visible and invisible work. *Computer Supported Cooperative Work (CSCW), 8*, 9–30.

Stember, M. (1991). Advancing the social sciences through the interdisciplinary enterprise. *The Social Science Journal, 28*(1), 1–14.

Tapscott, D., & Williams, A. D. (2006). *Wikinomics: How mass collaboration changes everything*. New York, NY: Portfolio.

Tekic, A., & Willoughby, K. (2018). Co-creation–child, sibling or adopted cousin of open innovation? *Innovation, 21*(2), 1–24.

Toffler, A. (1980). *The third wave*. New York: Bantam books.

Vargo, S. L., & Lusch, R. F. (2004). Evolving to a new dominant logic for marketing. *Journal of Marketing, 68*(1), 1–17. doi:10.1509/jmkg.68.1.1.24036

Vargo, S. L., & Lusch, R. F. (2006). Service-dominant logic: What it is, what it is not, what it might be. In R. F. Lusch, & S. L. Vargo (Eds.), *The service-dominant logic of marketing: Dialog, debate and directions* (pp. 43–56). Armonk: M.E. Sharpe.

Vargo, S. L., & Lusch, R. F. (2008a). Service-dominant logic: Continuing the evolution. *Journal of the Academy of Marketing Science*, *36*(1), 1–10.

Vargo, S. L., & Lusch, R. F. (2008b). From goods to service(s): Divergences and convergences of logics. *Industrial Marketing Management*, *37*, 254–259.

Velamuri, V. K., Neyer, A. K., & Möslein, K. M. (2011). Hybrid value creation: A systematic review of an evolving research area. *Journal für Betriebswirtschaft*, *61*(1), 3–35.

Voorberg, W. H., Bekkers, V. J. J. M., & Tummers, L. G. (2013). *Embarking on the social innovation journey: A systematic review regarding the potential of co-creation with citizens.* International Research Society for Public Management (IRSPM). Retrieved from http://hdl.handle.net/1765/39573

Voorberg, W. H., Bekkers, V. J., & Tummers, L. G. (2015). A systematic review of co-creation and co-production: Embarking on the social innovation journey. *Public Management Review*, *17*(9), 1333–1357.

Wonderful Copenhagen. (2017). *The end of tourism as we know it.* Copenhagen: Wonderful Copenhagen. Retrieved from http://localhood.wonderfulcopenhagen.dk/wonderful-copenhagen-strategy-2020.pdf

Wildlife tourism through the co-creation lens

Giovanna Bertella, M. Fumagalli and V. Williams-Grey

ABSTRACT
This study reflects on the conceptualisation of wild animals as co-creators. Its purpose is to encourage reflection about the role of animals in wildlife tourism. Therefore, to this end – and in the belief that diversity and creativity are important elements in critical thinking – the study was developed by a research team with diverse professional backgrounds. It adopts a fictional methodological approach, employing a fictive dialogue between a tourist joining a swim-with-dolphins tour and a dolphin and draws upon recent scholarly contributions on animals from the perspective of various disciplines, including philosophy, biology and tourism, The study's most important contribution comes in the form of a discussion of the co-creation concept from a critical perspective, based on innovative and explicitly-described ontological, epistemological and methodological considerations.

Introduction

The aim of this study is to drive further reflection on the adoption of the co-creation lens to wildlife tourism. The starting point is the observation that the suffix 'co-', that originates from Latin (*cum* = with), is usually used to form words whose meanings emphasise the concepts of togetherness, mutuality and in some cases also similarity, partnership and equality. Consequently, in this study, co-creation is regarded as a joint process involving at least two active actors who may be interested in being partners in the creation process. Inspired by a 'more than human' approach and, in particular, by ecofeminism (Haraway, 2007; Mies & Shiva, 2014), this study poses the following primary question: How and to what extent may wild animals be viewed as co-creators? Such question is relevant, given recent advancements in the tourism literature toward a recognition of the rights of nature – understood both holistically and at the level of individual organisms – and the related challenges around humans attempting to understand the perspective of non-human actors and, eventually, to speak for them (Dredge, 2018).

Ind and Coates (2013) observe that the co-creation concept has been adopted by academia in different ways. Campos, Mendes, Oom do Valle, and Scott (2018) identify two overall perspectives on co-creation in the tourism literature: a tourist perspective and a supply perspective. Within the first perspective, the experiential value is the focus and co-creation is understood mainly as a form of fruitful interaction. Within the second perspective, a management and marketing approach is adopted and knowledge and collaboration at the organisational and destination level are identified as the key factors for tourism development and management. In line with these perspectives, this study investigates two meanings of the co-creation concept: 'experiential value co-creation' through interaction and 'knowledge co-creation' through collaboration. Both perspectives are understood in terms of the third vector identified by Phi and Dredge in the editorial of this special issue, namely: critical issues in tourism co-creation.

Experiential value can be described as the tourists' perception of some functional, emotional, social, and epistemic value deriving from their experience and evolving in an idiosyncratic and dynamic way (Prebensen, Woo, & Uysal, 2014; Prebensen, Chen, & Uysal, 2017). Some studies associate the concept of experiential value to reciprocity and explore the possibility that tourist-guest relations might facilitate the emergence of value for both parties (Bertella, Cavicchi, & Bentini, 2018). This study aims to go a step further: namely, to broaden the concept of experiential value, including the perspective of the wild animals with whom the tourists interact.

In wildlife tourism, the typical values discussed in the literature are socio-economic benefits for the local communities; satisfaction, psychological and emotional benefits for the tourists, and educational outcomes

(Ballantyne, Packer, & Sutherland, 2011; Curtin, 2009; Higginbottom, 2004; Newsome, Dowling, & Moore, 2005). Benefits for the animals are usually included at the species level, in terms of conservation and protection, while possible risks, such as injuries and disturbance, are also referenced at the individual level (Higginbottom, 2004). Experiential value for the animals themselves seems to be overlooked by tourism scholars, something that might depend on a fundamental human-centric view of tourism.

In the light of the above-mentioned considerations, the following sub-question is raised:

> Sub-question 1: How can the experiential value that might emerge in wildlife tourism encounters be understood from the perspective of the animals?

With regard to the second meaning of co-creation explored in this study: namely, co-creation in terms of joint processes of tourism development and management; various types of actors who can affect – or are affected by – tourism may be identified. Relevant actors are usually identified amongst the following: tourism providers, tourists, public agencies at different levels, academia, host communities and NGOs. In the case of wildlife tourism, the critical aspect is the animal's role: animals are obviously affected by tourism, but to what extent can they affect it and, more specifically, its development and management? It can be assumed that if the animals could affect or influence tourism development and management, wildlife tourism might look very different. Presumably, this would apply to tourist activities which are lethal (e.g. trophy hunting), sub-lethal (e.g. catch-and-release fishing) and non-lethal (e.g. visiting zoos and dolphinaria). As suggested by the environmental ethics literature, the inclusion of wildlife among the tourism stakeholders is problematic, due to our difficulties and maybe unwillingness to fully understand, and eventually represent, the animals' interests (Holden, 2003).

This study raises a second sub-question:

> Sub-question 2: To what extent can humans and wild animals collaborate to create knowledge relevant to wildlife tourism management?

The article begins by presenting the basic ontological and epistemological assumptions for discussing issues relating to animals. The next section considers recent scholarly reflections on the use of animals in tourism, focusing in particular on the few studies that adopt the co-creation concept. The section on methodology describes the fictional approach used in this study to explore a particular case, i.e. swim-with-dolphins activities (i.e. tours offered by commercial operators to paying customers seeking in-water interactions with wild dolphins). This is followed by a fictitious dialogue between a dolphin and a tourist, preceded by a description of the commercial swim-with-dolphins sector. Finally, the insights gained from the dialogue are discussed and conclusions are drawn, including reflections on the theoretical contributions of this study and the methodological challenges of researching the animal world.

Theoretical background

There are two underlying assumptions that must be made explicit in order to discuss the role of animals in tourism with reference to the co-creation concept.

First assumption: what/who animals are

Recently, there has been a shift in the way animals are conceptualised in Western society and this can be noted in the increasing scrutiny of this issue by scholars from various disciplines, such as sociology, anthropology and philosophy (Kalof & Fitzgerald, 2007). Such conceptualizations derive from the application of various animal ethics approaches (Gruen, 2011). The common feature of these approaches is the rejection of a simplistic understanding of animals as undifferentiated objects, as in, for example, the Cartesian view of animals as *automata*: i.e. living beings that cannot feel pain/pleasure and do not have reasoning skills.

This study relies on the approach of ecofeminism (Mies & Shiva, 2014) and, in particular, those studies which postulate that animals are capable, both cognitively and emotionally, and human–animal relations can be meaningful for both parties (Gaard, 1993). Ecofeminists highlight the possible peculiarities of each species and individual and in this context, power relations in favour of humans are particularly critically reviewed and contrasted with attitudes and behaviour demonstrating respect and care (Adams & Gruen, 2014; Donovan & Adams, 2007; Gruen, 2015).

The ecofeminist perspective forms the basis of this study. According to this position, wild animals involved in tourism activities are sentient beings that can have complex and rich lives: they have intrinsic value and they can meaningfully interact with humans. This naturally leads us to a second assumption.

Second assumption: our knowledge about the animal world

Another relevant assumption refers to our potential to 'know' the animal world. In his 2016 book *Are we smart enough to know how smart animals are?*, the biologist,

Frans De Waal, notes that several experiments demonstrate that inter-species communication can occur (De Waal, 2016). At the same time, he reports that it has long been clear to biologists that each animal perceives the environment in his/her own way and this perception can compromise the possibility of reciprocal understanding and communication between different species. In this context, De Waal introduces the term 'anthropodenial', defined as 'the *a priori* rejection of human-like traits in other animals or animal-like traits in us'. He argues that anthropodenial can be a barrier to our knowledge of the animal world and a better approach could be to recognise that, the closer a species is to humans, the greater the chances of some reciprocal understanding and communication.

This study is predicated upon the belief that, to a certain degree, inter-species understanding and communication can occur on the basis of the common traits that we might share. In the case of wild animals, this can include for example: sharing evolutionary origins, physiology (mammals) and sociality traits (parental care, organisation in societies, etc.). The example of cetaceans (whales, dolphins and porpoises) is particularly interesting due to the scientific recognition of cetacean 'culture' as an important determinant of their highly-developed social behaviour (Kalof & Fitzgerald, 2007).

Our possibility to understand the animal world can be related to the ecofeminist concept of 'entangled empathy', i.e. a caring perception based on our efforts to understand and attend to the animals' needs, interests, desires, vulnerabilities and sensitivities (Gruen, 2015). Entangled empathy is a relational type of attention based on our cognition and the emotions triggered by our interactions with animals. In line with this position, this study argues that our knowledge of animals can derive from connecting with them through our cognition, as well as our emotions.

Animals and co-creation in tourism

In 2009, a special issue of *Current Issues in Tourism* was published: nine of the articles published in the issue discussed and elaborated conceptual positions related to the relationship between humans and animals in tourism, suggesting management strategies useful to ensure the quality of the experiences for the tourists, whilst ensuring the animals' welfare and rights (Carr, 2009).

Since the publication of that issue and the book *Tourism and Animal Ethics* (Fennell, 2012), the tourism literature on animals, both domesticated and wild, has increased considerably. Several tourism publications have adopted a critical stance on the inclusion of the animals in the tourism industry, and some of them adopting the ecofeminism approach (Bertella, 2018; Yudina & Fennell, 2013; Yudina & Grimwood, 2016).

To the authors' knowledge, the adoption of the co-creation lens to wildlife tourism has not yet been specifically discussed. Writing about value co-creation, Bertella (2014) analyses the dog sledding experience, recognising that dogs, as subjects, play a crucial role in the emergence of experiential value for the tourists. In her study, the human–animal interactions are described as reciprocal, given that the animal species involved – dogs – in addition to being domesticated, are selected by the tourism provider because they are particularly friendly to humans. In this context, the author reports the emergence of value also for the animals, who can have friendly encounters with the tourists and be physically active during non-competitive trips. Nonetheless, no particular focus on the animals' perspective is included, with the result that the predominant perspective is that of the tourists. In her conclusions, Bertella (2014) reflects on the methodological challenges of researching animals and their relationship with humans: an aspect particularly relevant to co-creation. The researcher's respect and empathy for animals are critical factors required to overcome such challenges alongside his/her competence and familiarity with the focal species, as well as with individual animals.

Campos, Mendes, Oom do Valle, and Scott (2017) also apply the co-creation lens to animal-based tourism. They investigate dolphins in captivity and include activities such as swimming and playing with these animals. This study explicitly adopts the perspective of the tourists, focusing on on-site co-creation, understood as the process through which the tourists' subjectively lived experience and the related value evolve. The authors discuss neither the underlying understanding of the animals, nor the human–animal interactions and the related ontological and epistemological aspects.

With regard to tourism knowledge and management and the role of animals as actively engaged and influential actors, it is hard to find in the tourism literature any contribution that adopts the co-creation concept. Some tourism studies discuss how animals as stakeholders may benefit from tourism. Some examples specifically relating to cetacean tourism are cited by Higham, Bejder, and Williams (2014). Nonetheless, no scholar has ever considered those aspects through the co-creation lens.

Methodology

The historical development of qualitative inquiry is extensively discussed in the literature and several scholars note a recent shift toward a more diffused adoption and acceptance of multiple modes of understanding (Bramwell & Lane, 2000; Denzin & Lincoln, 1994; Jamal & Hollinshead, 2001; Moses & Knutsen, 2007; Riley & Love, 2000; Wilson & Hollinshead, 2015). This shift implies collaboration across disciplines and among all relevant stakeholders, and includes ontological, epistemological and methodological positions that take into consideration the perspectives of traditionally 'voiceless' actors, such as various minorities.

One such alternative means of understanding, Creative Analytic Practice (CAP), includes various methods of expression such as fiction and poetry and is viewed as a potentially fruitful way to imagine and reflect on the complexity of lived experiences (Parry & Johnson, 2007; Richardson & St. Pierre, 2005). In order to be useful as a method of inquiry, authors using CAP are expected to be particularly open and critical in their role (reflexivity), and their texts are expected to satisfy some criteria such as: aesthetic merit (be engaging), impact (generate curiosity and new questions), and reality (present a credible account of a situation).

When considering the shift in qualitative inquiry toward a more diffused adoption and acceptance of multiple modes of understanding, some limitations can be identified in the tourism literature relating to animals presented in the previous section. Although some studies contribute to alternative ways to conceptualise animals, relevant epistemological and methodological issues are not discussed in detail. The methodological approaches adopted are also quite limited in relation to possible ways to explore the animal's perspective. Moreover, the two studies utilising the co-creation concept (Bertella, 2014; Campos et al., 2017) do not include among the authors any scholar from relevant natural sciences (e.g. zoology), whose perspectives and knowledge could help achieve a deeper understanding and investigation of animals in tourism.

In order to meet those challenges, this study utilises a research team comprising individuals with different professional backgrounds, and uses fiction with the intention of enlarging our understanding of a particular situation (swim-with-dolphin tours) including the perspective of a usually 'voiceless' non-human actor (a dolphin).

The research team

Our team includes individuals with different professional backgrounds: two academics (PhD: business management and tourism; PhD: zoology) and a researcher/campaigner at an international non-profit organisation dedicated to the conservation and welfare of cetaceans. The team members are passionate about the animal world and have extensive practical experience in wildlife research and tourism, including fieldwork and engagement with all relevant stakeholders (Table 1). The team's background may be described as a combination of theoretical and practical competence in business, tourism, biology, marine ecology, environmental and wildlife conservation, interpretation and communication.

Fiction and the construction of a human–dolphin dialogue

Several scholars from the field of social sciences argue that complex phenomena can be investigated through the adoption of fiction (Banks & Banks, 1998; Gough, 2008; Reinhold, 2018). Fiction can help us to go beyond reality and explore phenomena in a deeper

Table 1. The research team's practical knowledge and experience in relation to wildlife tourism operators and tourists, and dolphins.

	Practical knowledge and experience
Wildlife tourism operators: their knowledge about wildlife, their attitude and behaviour toward wildlife, how they design and manage wildlife experiences	Internationally gained knowledge and experience. Presenting at training workshops for whale watching operators; interviewing operators on what they need in order to make their business more conservation oriented and responsible; attendance at workshops and seminars and participation in round-table discussions with operators on all aspects of their industry including passenger safety, satisfaction and the risks/concerns around the swim industry; working for commercial operators (as guides).
Wildlife tourists: their knowledge and perception of wildlife, their motivation, what they look for when engaging in wildlife encounters, what they value and remember	Internationally gained knowledge and experience. Volunteering for NGOs: tasks in direct contact with tourists pre, during and post wildlife trips (presentations at hotels, information centres, beach patrols); designing and delivering tourism surveys to whale watch passengers; working as guide on a whale watching boat; answering passenger questions including about swimming with cetaceans; presentations and interpretation to tourists (both onboard and onshore before and after trips); interviewing whale watching passengers (satisfaction surveys, etc.); providing information to passengers on what to look for in a good whale watching trip; informing tourists through articles, reports, flyers, online articles and blogs, about responsible whale watching and swim-with-dolphins tours, including related dangers; personal experience as wildlife tourists.
Dolphins: their behaviour in natural settings and in relation to human presence	Tasks in direct contact with dolphins: fieldwork as part of volunteering activities, PhD education and research.

way, replacing or integrating those more traditional inquiry modes that rely exclusively on reason Fictional stories have several strengths: they can ably represent the complexity and the particularity of a situation; increase the variety of questions that we ask; engage readers and engender empathy (Eisner, 1997).

Furthermore, fictional writing can be used to express nuances and explore concepts that can be difficult or impossible to grasp using more traditional research methods (Brunila & Valero, 2018). Following the CAP criteria of aesthetic merit, impact and reality, writing about imagined events can allow both authors and readers to explore rarely-debated issues from new perspectives.

In the tourism literature, there is increasing interest in the opportunity to use creativity and, in particular, fiction in research inquiries (Phillimore & Goodson, 2004; Wilson & Hollinshead, 2015). Recently, a fictional narrative has been applied to tourism research in relation to the possible representation of future scenarios (Yeoman & Postma, 2014). In the case of wildlife tourism, Wright (2018) adopts science fictional narrative to explore future developments, with reference to the use of cloned animals.

Within this context, creative writing plays an important role. It is interesting to note that creative writing is sometimes commented on with reference to our 'animality' (Reinhold, 2018). Haraway (2007), McHugh (2011) and Bell, Instone, and Mee (2018) reflect on some methods that might be adopted to investigate the animal world, whilst attempting to take the animals' perspective.

Based on such considerations, this study develops a fictitious tourist-dolphin dialogue occurring during an organised swim-with-dolphins tour in a wild setting. The choice to elaborate a dialogue relies on the potential of screenplays to invite the readers to new interpretations (Berbary, 2011). Moreover, the dialogue approach is inspired by Plato's Socratic dialogues, with particular emphasis on the inquisitive attitude of a character – the dolphin – in an attempt to encourage further reflection in the other character – the tourist.

The construction of the fictitious dialogue

Rows 2 and 3 in Table 1 present an overview of the authors' practical knowledge and experience that, in addition to their theoretical knowledge, were relevant to the elaboration of the dialogue: in particular, the roles of the tourist and the dolphin.

In addition, insights into the tourist's perspective were gained through consulting scientific literature, in particular: Curtin (2006, 2009) and DeMares and Krycka (1998), as well as grey literature; in particular: posts on TripAdvisor. With regard to the latter, a search using the expression 'swim/swimming with dolphins' was conducted and the company with the most reviews (190) was selected. The vast majority of the reviews (179) were very positive, with only 9 being negative. The 20 most recent positive reviews and the 9 negative reviews were analysed in order to identify which aspects of the experience were most commented upon by tourists.

In terms of 'voicing' the dolphin, the authors relied on their practical experience (Table 1, row 3) as well as their creativity and some inspirational sources. The latter were the character Spock from the 1986 movie *Star Trek IV: The Voyage Home* (in particular, his comments on human arrogance) and the 1972 novel *Watership Down*. The latter was used as a model to guide the style and tone of the language spoken by the dolphin.

Co-creation in swim-with-dolphins

The context where the dialogue occurs

> 'Swim-with-dolphins' is a general term used to define any commercial activity offering paying customers in-water interactions with wild dolphins in wild settings (Parsons et al., 2006). This type of tourism has increased in popularity over the last few years (Convention on Migratory Species, 2017; Hoyt, 2001).

In most cases, interactions are pursued by humans through invasive approaches (*active* interaction) or initiated by dolphins of their own accord (*passive* interaction), and less often, they are solicited with food provisioning (Parsons et al., 2006; Samuels, Bejder, Constantine, & Heinrich, 2003).

While swim-with-dolphins operations inevitably differ from one context to another, the essence of the experience, as well as its associated benefits and risks, have some features which might be considered similar to any wildlife tourism encounter. However, humans swimming in proximity to large, powerful marine mammals could also be at significant risk of harm, injury or death, and also the potential risk of disease transmission (Shane, Tepley, & Costello, 1993; Waltzek, Cortés-Hinojosa, Wellehan, & Gray, 2012).

Dolphins respond to close approaches by changing their behaviour and movements, and likely a suite of other responses as field studies are mainly observational (Fumagalli et al., 2018; Machernis, Powell, Engleby, & Spradlin, 2018). Such responses are not only species-specific, but can vary within species and populations, and even the same individual dolphin may show a different inclination to engage in, or sustain, an in-water interaction at different times or life stages (Convention on Migratory Species, 2017).

The simple act of responding to an interaction (whether this is to engage in it or to avoid it) poses an additional demand and can upset the animal's energetic balance, with detrimental effects on survival or reproduction rates (Bejder et al., 2006; Lusseau, 2004). In the longer term, a population exposed to interactions may be at risk of decline, or displacement to less disturbed sites. In order to preserve these wild populations, precaution is recommended in the sanctioning or management of these interactions, especially when swim-with-dolphin tours target critical habitats, species with particular spatial or temporal constraints on their behaviour, or already-threatened species or populations (Convention on Migratory Species, 2017).

With regard to regulations, the swim-with-dolphins industry is managed and regulated in different ways around the globe. In some regions, the activity is prohibited outright (e.g. the Canary Islands), whilst in others (e.g. New Zealand) it is legal and regulated. Elsewhere, this type of activity lacks any kind of formal regulation and relies on national or regional guidelines and codes of conduct outlining best practices. These latter approaches may be issued by governments or official agencies, or by the industry itself as, for example, in Japan or Iceland. In other regions, however, the activity may be entirely unmanaged, even lacking informal codes of conduct or advice and these scenarios obviously prompt the greatest concern for the safety and welfare of all stakeholders.

The fictitious dialogue

As strange as it may seem, every now and then, quite inexplicably, inter-species communication occurs. This happened to a tourist joining a swim-with-dolphins tour. The content of the dialogue is reported below.

Tourist: Hello … I'm so excited to meet you!

Dolphin: Hello to you! What are you doing here? Looking for food? Are you injured? Why are you moving so strangely?

Tourist: What?! I thought it was just beautiful to see you swimming, all your splashy and elegant movements … and now I can even understand you! And you understand me! This is so awesome! I'm not injured … it's just that I'm not such a good swimmer as you are!

Dolphin: I see that … you are indeed rather clumsy! So, what are you doing here?

Tourist: I'm here to see you.

Dolphin: I didn't expect your visit, but since you are here … Have we met before? I can't remember … I have met others like you before though.

Tourist: You mean other humans …

Dolphin: We call you *Trizzbz.* It means something like 'those-who-move-in-and-out-of-water'.

Tourist: I'm here because I like others like you, we call you dolphins.

Dolphin: Oh … so I'm a dolphin!

Tourist: What do you call yourselves? I mean your … species … your group … ?

Dolphin: *We.*

Tourist: It makes sense … It's amazing … you have names for us humans and for yourselves … it's like a language, like we have … I'm so impressed!

Dolphin: You know, you *Trizzbz* might not be so special and unique as you think. You are just one type of life among many others. Only arrogance can put a type of life in the centre and all the rest around … like ripples in the water, when real life is made of circles, waves, eddies – and sand and rocks and *dbgjd*, and many *jsfdjs* …

Tourist: I get your point. I guess that sometimes we *Trizzbz* are indeed quite … arrogant.

Dolphin: So, what do you like about seeing *We*?

Tourist: You are so cute! You look very happy … and free.

Dolphin: Happy … I guess so … happiness is being healthy, isn't it?

Tourist: More than that actually … it's … um … all about enjoying your time and not worrying.

Dolphin: I worry a lot … the fights, the food to search for, the young ones, the diseases, all the dangers out there … it is so exhausting … so … I guess I'm not as happy as you think …

Tourist: Oh, I see … Our tour guide told us a bit about your life in the ocean. It's not easy to live in the wild … it can be hard to find food and avoid injuries … But when you approached me, you looked like you were smiling!

Dolphin: I'm not sure I understand what you mean. This is just my face. I don't know what you mean by smile. The fact is that I'm curious … you know? I wanted to see what you were. They say that I'm the most curious in my family, that's also why I got my name!

Tourist: Do you have a name?

Dolphin: Of course I have a name! My name is *Blutsdn*, which means something like 'the-daughter-of-*Blutsjin*-who-is-always-looking-for-something'.

Tourist: So interesting!

Dolphin: They have warned me though … about *Trizzbz.* Some are so pushy! They come too close while *We* are busy doing other things, you know … important things like feeding or sleeping! They can distract our young ones. And then there are those stories about … abduction … and also … murders …

Tourist: This is sad. Sorry about that, some … *Trizzbz* can be dangerous. I'm actually here with other humans now … they are over there, some are in the water and some are still on the boat ... I don't know them personally, I hope none of them are dangerous ... pushy, you know ... I'm quite sure none of them are .... dolphin murderers ... just maybe a bit over-enthusiastic ...

Dolphin: You don't know the members of your own family?

Tourist: I'm not here with my family! The others ... well, we just met ... we don't know each other.

Dolphin: I see ... I'll keep this in mind. We've had a tough day today – can you stop them from getting in the water?

Tourist: ... I could try ... but I don't think they will listen to me. Also, I don't know how the other humans may behave once in the water. Maybe they will be like me?

Dolphin: Maybe. But *Trizzbz* are unpredictable, *We* often say that.

Tourist: Not me, I'm very predictable ... please stay a little longer ...

Dolphin: You look clumsy rather than dangerous, so I'll hang around a little longer. So, do you still like to see us, even though you now know that we are not always happy?

Tourist: Mmmhh ... I must admit that this does sound a bit strange. To me, you look happy, but of course if you say that you are often worried, perhaps you are not so happy after all ...

Dolphin: It's just that there is always so much to do ...

Tourist: Yes, but, you know, this is such a beautiful place ... and you are free ...

Dolphin: You mean free ... to worry?

Tourist: No, I mean free to ... move, swim and jump. And the noise you make ... your voice sounds so exciting, like children laughing!

Dolphin: Aren't you free to move? I see that in the water you are quite a disaster when it comes to movements, but maybe out of the water ... ?

Tourist: Oh yes, we are free to move, but it is different ... we have to work ...

Dolphin: Work? Is it a kind of worry?

Tourist: Well ... you could say so, it is something that we usually must do to support ourselves and our families. It takes a lot of our time and energy and it is sometimes very stressful. It's not something we always choose to do or are happy to do.

Dolphin: *We* do many things and some are ... you know, not a real choice. But, I have to admit, it never crosses my mind that getting out of the water and looking at *Trizzbz* could help! Whereas you think that coming here to see us will make you feel better, don't you? You come to have a look at us because you think that we have something you don't have and this can make you feel happy and free too, right?

Tourist: Mmmmhh ... I think that, yes, you could put it like that. Maybe, we humans tend to see what we want to see, in order to feel better?

Dolphin: It sounds a bit strange to me ... however, for me, it is ok to meet you and other *Trizzbz*, as long as I am not too tired or busy ... But not the dangerous ones, those I don't want to see ...

Tourist: The pushy ones and the other ones ... yes, try to stay as far as possible from them!

Dolphin: There are different types of *Trizzbz* ... this we have learned! *We* can also be very different. You'll never meet old *Blutsvpdn*, he is so wary!

Tourist: Doesn't he like to meet *Trizzbz*?

Dolphin: It has nothing to do with *Trizzbz*, you know. That's the way he is. And I hope you never meet *Bfkjsofjis* ... he is a troublemaker, quite aggressive ... Oh! I have to go, they are calling me ...

Tourist: Really? I didn't notice, are they calling your name? And what ... wait, wait! ... can't you stay a bit longer?

Discussion of the sub-questions

Sub-question 1: How can the experiential value that might emerge in wildlife tourism encounters be understood from the perspective of the animals?

This study suggests that this value can be related to the animal's degree of curiosity. At the same time, it highlights the possible differences between individual animals: some might be particularly shy or wary, so in some instances, the experience would be lacking in value and the presence of humans might be experienced only as an annoyance. Other animals might regard it as a threat, due to some previous negative encounter with humans, whether direct or indirect. In well-managed swim-with-dolphin tours for example, the dolphins might have the opportunity to come closer or to move away, but human presence may nonetheless be perceived as intrusive. As reported by *Blutsdn*, no dolphin has ever tried to emerge from the water and impose her/his presence to humans! Curious animals might approach swimmers or poke their head above the surface to look at the boats and people around them. This type of intentional contact is, however, far less invasive than occupying a space in someone else's world, even temporarily. Imagine the sudden appearance of a dolphin in our living room, trying to engage our attention whilst we are busy with our daily life!

It can be noted that *Blutsdn* does not ask the tourist his name. This might indicate that, although dolphins are aware of individual differences among humans, *Blutsdn*, might tend to think about us in collective terms. This implies that the type of relation necessary for the emergence of value in terms of 'friendship' might not be interesting to wild animals. Moreover, the fact that the dolphins refer to themselves as *We* may suggest a way of reasoning in terms of We and 'Others'.

Blutsdn tells the tourist about *Bfkjsofjis*, a dolphin described as a troublemaker, who might initiate contact with humans with the intent to cause mischief. For some animals, the encounter with a human might trigger the emergence of this type of value that is almost certainly not reciprocal.

With reference again to the reciprocal aspect of the experiential value deriving from the human–dolphin encounter, the dialogue also highlights how this should be carefully considered with regard to an important aspect raised by *Blutsdn*: namely, the possible different behaviours amongst tourists. Even the most responsible tourism provider has only limited control over their clients' behaviour in the water. While some tourists might do their best to create a mutual type of experiential value, others might have a very different attitude, either based upon ignorance or their personality. Troublemakers exist among humans as well as among dolphins.

Further considerations are necessary with regard to the first sub-question relating to the role of wild animals in the value-creation process. The dialogue suggests that *Blutsdn*'s understanding of her role in such a process is very limited. *Blutsdn* appears not to fully understand that the reason the tourist is in the water – an element where he does not feel particularly confident – is to observe the dolphins, based on the perceived belief that they are 'happy'. This latter aspect is relevant also in relation to the second sub-question.

Sub-question 2: To what extent can humans and wild animals collaborate to create knowledge relevant to wildlife tourism management?

The dialogue is built on the fictional possibility for humans and dolphins to communicate relatively easily with each other. Nonetheless, some elements suggest that communication is actually somewhat difficult. Some terms and concepts can be translated, but not all. In the same way, we are able to understand some dolphin behaviour (and likely they understand some of ours), but we are far from deciphering each other's full repertoire, intentions and language. For example, what does *Blutsdn* mean when she says that real life is made of *dbgjd* and *jsfdjs*? Also, the human world remains somewhat obscure to the dolphin: why are humans happy when observing dolphins struggling to survive in the wild? Aren't health and happiness the same? Are humans free in their world? Why are humans so surprised that dolphins have names? Many questions about humans remain unanswered for *Blutsdn* who, in the end, returns to her life in the dolphin realm. Similarly, some questions remain partially unanswered for the tourist and presumably also the reader.

Based on these reflections, we propose that tourist-wild dolphin relationships in the context of commercial swim-with-dolphin tours are necessarily restricted to occasional encounters and cannot develop into any real, long-lasting collaboration.

Conclusions

This study posed a primary question (How and to what extent may wild animals be viewed as co-creators?) and two sub-questions (How can the experiential value that might emerge in wildlife tourism encounters be understood from the perspective of the animals? To what extent can humans and wild animals collaborate to create knowledge relevant to wildlife tourism management?). Based on the discussion of the two sub-questions, some conclusions may be drawn regarding the main research question, namely the possibility of viewing wildlife tourism through the co-creation lens. Although reciprocal experiential value can emerge – mainly in the form of curiosity – when considering the overall experience, this may be just one of many possibilities, including the potential for unbalanced encounters, whereby only one party experiences value. Moreover, although they are key actors, it can be assumed that wild animals have limited knowledge of the human world, and are ignorant of what tourism as an important sector of many human economies is and their role in some tourism encounters such as wildlife watching tours. This considerably limits their ability to participate in knowledge co-creation relevant to tourism management.

Our conclusion is that the possibility of viewing wild animals as co-creators is problematic on several levels. This study suggests that the required understanding and communication processes upon which reciprocal experiential value and collaboration should be based, are hard to obtain. Additionally, problems may not be limited to issues around understanding and communication, as there might also be a lack of interest in interacting and collaborating with humans on the animals' part.

This study's theoretical contribution may be viewed in relation to the two research papers previously identified in the tourism literature relating to co-creation and animals. Campos *et al.* (2017) apply this concept to the case of wild animals but do not discuss this in any detail. By contrast, this study starts with the etymological meaning of the term, discusses several relevant aspects and concludes that its use might be not particularly suitable. Bertella (2014) also applies the co-creation concept to animals; but in this instance, her case study is limited to sled dogs. The challenges outlined in the Bertella paper suggest that, whilst the use of the co-creation lens might be reasonable in the case of domesticated animals, a quite different approach may be required in the case of wildlife. We suggest that the concept of 'entangled empathy' might offer a valuable alternative approach to wildlife tourism; in particular, tourism encounters and their management might usefully be adopted as a starting point for the importance of respecting individual differences between animals, rather than an idealised human-wildlife relationship.

This study's methodological contribution concerns the use of CAP to investigate animals in tourism. If fully embraced, the recognition of the richness of the animal mind and world has important implications in terms of research methodology. There is a clear gap in the literature in this sense and CAP might offer a viable means of plugging such a gap. The authors found it quite challenging to develop the fictional human–dolphin dialogue but were aided by sharing a perspective on animals and the complexity of their inclusion in tourism. The team also acknowledges the importance of being open with each other, as well as with the reader, about their varying, albeit complementary, backgrounds and experience. In general, the team found the suggestions made by Parry and Johnson (2007) regarding CAP quality criteria – in particular as regards reflexivity – to be useful. The major challenge has been in envisaging and plausibly interpreting the dolphin's thoughts and emotions. In this regard, we feel that Parry and Johnson's use of the term 'reality' can be misleading and could usefully be replaced by a criterion of 'plausibility'.

Despite the possible limitations of the empirical part (a fictitious case and a dialogue limited to two actors), this study contributes to the wildlife tourism literature on the challenging topic of human–wild animal relations, adopting an ecofeminist perspective and a creative methodology. Moreover, this study contributes towards bridging the literature gap concerning the adoption of the co-creation concept to wildlife tourism. For future studies, we consider the following as important prerequisites for researching co-creation in human–animal encounters: theoretical and practical knowledge about the non-human actors (at the species level and ideally at the individual level), interdisciplinary research and a genuine interest and commitment to respect the animals.

Our fictional narrative is a novel and, hopefully, effective first step towards engaging others to approach the animal world and critically consider the way we (scholars, operators, legislators, tourists, etc.) enter it. Thus, this study is an invitation to explore alternative ways to view, frame and, finally, to manage wildlife tourism.

Disclosure statement

No potential conflict of interest was reported by the authors.

ORCID

Giovanna Bertella http://orcid.org/0000-0001-5530-8588
M. Fumagalli http://orcid.org/0000-0003-2954-5736

References

Adams, C., & Gruen, L. (2014). *Ecofeminism. Feminist intersections with other animals and the Earth*. New York, NY: Bloomsbury.

Banks, A., & Banks, S. P. (1998). *Fiction and social research: By ice or fire*. Walnut Creek: AltaMira Press.

Ballantyne, R., Packer, J., & Sutherland, L. A. (2011). Visitors' memories of wildlife tourism: Implications for the design of powerful interpretive experiences. *Tourism Management, 32* (4), 770–779.

Bejder, L., Samuels, A., Whitehead, H., Gales, N., Mann, J., Connor, R., … Kruetzen, M. (2006). Decline in relative abundance of bottlenose dolphins exposed to long-term disturbance. *Conservation Biology, 20*(6), 1791–1798.

Bell, S. J., Instone, L., & Mee, K. J. (2018). Engaged witnessing: Researching with the more-than-human. *Area, 50*(1), 136–144.

Berbary, L. A. (2011). Poststructural writerly representation: Screenplay as creative analytic practice. *Qualitative Inquiry, 17*(2), 186–196.

Bertella, G. (2014). The co-creation of animal-based tourism experience. *Tourism Recreation Research, 39*(1), 115–125.

Bertella, G., Cavicchi, A., & Bentini, T. (2018). The reciprocal aspect of the experience value: Tourists and residents celebrating weddings in the rural village of Petritoli (Italy). *Anatolia, 29*(1), 52–62.

Bertella, G. (2018). Sustainability in wildlife tourism: Challenging the assumptions and imagining alternatives. *Tourism Review*. doi:10.1108/TR-11-2017-0166.

Bramwell, B., & Lane, B. (2000). *Tourism collaboration and partnerships: Politics practice and sustainability*. Clevedon: Channel View.

Brunila, K., & Valero, P. (2018). Anxiety and the making of research(ing) subjects in neoliberal academia. *Subjectivity, 11*(1), 74–89.

Campos, A. C., Mendes, J., Oom do Valle, P., & Scott, N. (2017). Co-creating animal-based tourist experiences: Attention, involvement and memorability. *Tourism Management, 63*, 100–114.

Campos, A. C., Mendes, J., Oom do Valle, P., & Scott, N. (2018). Co-creation of tourist experiences: A literature review. *Current Issues in Tourism, 21*(4), 369–400.

Carr, N. (2009). Animals in the tourism and leisure experience. *Current Issues in Tourism, 12*(5-6), 409–411.

Convention on Migratory Species. (2017). *Recreational in-water interaction with aquatic mammals*. Report prepared by the Aquatic Mammals Working Group of the Scientific Council. UNEP/CMS/COP12/Inf.13.

Curtin, S. (2006). Swimming with dolphins: A phenomenological exploration of tourist recollections. *International Journal of Tourism Research, 8*(4), 301–315.

Curtin, S. (2009). Wildlife tourism: The intangible, psychological benefits of human-wildlife encounters. *Current Issues in Tourism, 12*(5–6), 451–474.

DeMares, R., & Krycka, K. (1998). Wild animal triggered peak experiences: Transpersonal aspects. *Journal of Transpersonal Psychology, 30*, 161–177.

Denzin, N., & Lincoln, Y. (1994). *A handbook of qualitative research*. London: Sage.

De Waal, F. (2016). *Are we smart enough to know how smart animals are?* London: Norton.

Donovan, J., & Adams, C. J. (2007). *The feminist care tradition in animal ethics*. New York, NY: Columbia University Press.

Dredge, D. (2018). Governance, tourism and resilience: A long way to go? In J. Saarinen, & A. M. Gill (Eds.), *Resilient destinations and tourism. Governance strategies in the transition towards sustainability in tourism wildlife tourism* (pp. 56–93). London: Routledge.

Eisner, E. W. (1997). The promise and perils of alternative forms of data representation. *Educational Researcher, 26*(6), 4–10.

Fennell, D. (2012). *Tourism and animal ethics*. London: Routledge.

Fumagalli, M., Cesario, A., Costa, M., Harraway, J. A., Notarbartolo Di Sciara, G., & Slooten, E. (2018). Behavioural responses of spinner dolphins to human interactions. *Royal Society Open Science, 5*(4), 172044. DOI: 10.1098/rsos.172044.

Gaard, G. (1993). *Ecofeminism. Women, animals, nature*. Philadelphia, PA: Temple University Press.

Gough, N. (2008). Narrative experiments and imaginative inquiry. *South African Journal of Education, 28*, 335–349.

Gruen, L. (2011). *Ethics and animals. An introduction*. New York: Cambridge University Press.

Gruen, L. (2015). *Entangled empathy. An alternative ethic for our relationships with animals*. Brooklyn, NY: Lantern Books.

Haraway, D. (2007). *When species meet*. Minneapolis: University of Minnesota Press.

Higginbottom, K. (2004). *Wildlife tourism. Impacts, management and planning*. The Gold Coast: Common Ground Publishing Pty Ltd and Cooperative Research Centre for Sustainable Tourism.

Higham, J., Bejder, L., & Williams, R. (2014). *Whale-watching. Sustainable tourism and ecological management*. Cornwall: Cambridge University Press.

Holden, A. (2003). In need of new environmental ethics for tourism? *Annals of Tourism Research, 30*(1), 94–108.

Hoyt, E. (2001). *Whale watching 2001: Worldwide tourism numbers, expenditures and expanding socioeconomic benefits*. Crowborough: International Fund for Animal Welfare, UNEP.

Ind, N., & Coates, N. (2013). The meanings of co-creation. *European Business Review, 25*(1), 86–95.

Jamal, T., & Hollinshead, K. (2001). Tourism and the forbidden zone: The underserved power of qualitative inquiry. *Tourism Management, 22*, 63–82.

Kalof, L., & Fitzgerald, A. (2007). *The animals reader. The essential classic and contemporary writings*. Oxford: Berg.

Lusseau, D. (2004). The hidden cost of tourism: Detecting long-term effects of tourism using behavioral information. *Ecology and Society, 9*(1), 2.

Machernis, A. F., Powell, J. R., Engleby, L., & Spradlin, T. R. (2018). *An updated literature review examining the impacts of tourism on marine mammals over the last fifteen years (2000-2015) to inform research and management programs*. NOAA Technical Memorandum NMFS-SER-7. DOI: 10.7289/v5/tm-nmfs-ser-7.

Wright, D. W. M. (2018). Cloning animals for tourism in the year 2070. *Futures, 95*, 58–75.

McHugh, S. (2011). *Animal stories: Narrating across species lines*. Minneapolis: University of Minnesota Press.

Mies, M., & Shiva, V. (2014). *Ecofeminism*. London: Zed Books.

Moses, J., & Knutsen, T. (2007). *Ways of knowing: Competing methodologies in social and political research. Basingstoke*. Hants: Palgrave.

Newsome, D., Dowling, R. K., & Moore, S. A. (2005). *Wildlife tourism*. Clevedon: Channel View Books.

Parsons, E. C. M., Fortuna, C. M., Ritter, F., Rose, N. A., Simmonds, M. P., Weinrich, M., … Panigada, S. (2006). Glossary of whale watching terms. *Journal of Cetacean Research and Management, 8* (Supplement), 249–251.

Parry, D. C., & Johnson, C. W. (2007). Contextualizing leisure research to encompass complexity in lived leisure experience: The need for creative analytic practice. *Leisure Sciences, 29*(2), 119–130.

Phillimore, J., & Goodson, L. (2004). Progress in qualitative research in tourism. In J. Phillimore, & L. Goodson (Eds.), *Qualitative research in tourism: Ontologies, epistemologies, methodologies* (pp. 3–299). London: Routledge.

Prebensen, N., Woo, E., & Uysal, M. (2014). Experience value: Antecedents and consequences. *Current Issues in Tourism, 17*(10), 910–928.

Prebensen, N., Chen, J., & Uysal, M. (2017). *Co-creation in tourist experiences*. Oxon: Routledge.

Reinhold, E. (2018). How to become animal through writing: The case of the bear. *Culture and Organization*, *24*(4), 318–329.

Richardson, L., & St. Pierre, E. A. (2005). Writing: A method of inquiry. In N. K. Denzin, & Y. S. Lincoln (Eds.), *The Sage handbook of qualitative inquiry* (pp. 959–978). Thousand Oaks, CA: Sage.

Riley, R., & Love, L. (2000). The state of qualitative tourism research. *Annals of Tourism Research*, *27*, 164–187.

Samuels, A., Bejder, L., Constantine, R., & Heinrich, S. (2003). Swimming with wild cetaceans in the Southern Hemisphere. In N. Gales, M. Hindell, & R. Kirkwood (Eds.), *Marine mammals: Fisheries, tourism and management issues* (pp. 265–291). Collingwood: CSIRO.

Shane, S. H., Tepley, L., & Costello, L. (1993). Life threatening contact between a woman and a pilot whale captured on film. *Marine Mammal Science*, *9*, 331–336.

Waltzek, T. B., Cortés-Hinojosa, G., Wellehan Jr., J. F., & Gray, G. C. (2012). Marine mammal zoonoses: A review of disease manifestations. *Zoonoses and Public Health*, *59*(8), 521–535.

Wilson, E., & Hollinshead, K. (2015). Qualitative tourism research: Opportunities in the emergent soft sciences. *Annals of Tourism Research*, *54*, 30–47.

Yeoman, I., & Postma, A. (2014). Developing an ontological framework for tourism futures. *Tourism Recreation Research*, *39* (3), 299–304.

Yudina, O., & Fennell, D. (2013). Ecofeminism in the tourism context: A discussion of the use of other-than-human animals as food in tourism. *Tourism Recreation Research*, *38* (1), 55–69.

Yudina, O., & Grimwood, B. (2016). Situating the wildlife spectacle: Ecofeminism, representation, and polar bear tourism. *Journal of Sustainable Tourism*, *24*(5), 715–734.

Co-creating knowledge in tourism research using the Ketso method

Yana Wengel, Alison McIntosh and Cheryl Cockburn-Wootten

ABSTRACT
Tourism scholars have called for critical engagement with transformational co-creative methodologies. Within this call, there is a need for researchers to be positioned as facilitators and co-creators; rather than lone experts. We provide a critical review of the Ketso method. Ketso is a facilitated 'workshop in a bag'; a toolkit that enables people to think and work together. Ketso can be used for data collection and as a supplementary analysis tool. Critical reflections on Ketso are provided to illustrate how it co-creates knowledge and collaborative solutions for transformational tourism. As a data collection tool, Ketso provides an innovative and authentic approach to stakeholder collaboration and decision making. As a supplementary data analysis tool, it provides an opportunity to address some of the limitations of thematic analysis such as simplicity and lack of coherence. In providing critical reflections on Ketso, we contribute to future thinking for the adoption of this co-creative method for tourism research.

Introduction

Tourism scholars have called for a critical engagement with transformational research methodologies as they noted that tourism still seems enamoured with traditional methods of enquiry which continue to reproduce existing knowledge, activities and language practices (Gillovic, McIntosh, Darcy, & Cockburn-Wootten, 2018, p. 23; Sedgley, Pritchard, & Morgan, 2011). Tourism researchers have commented that '[d]espite this mass of research ... we are failing to answer questions', calling for scholars to think creatively to problems, critique assumptions, analyse rhetoric and evaluate the broader power discourses to gain a deeper understanding of relationships (Singh, 2012, p. 23). To address this failure, critical researchers have advocated for methodological activities that move beyond the 'academic as epidemic' approach to, instead, adopting collaborative frameworks that disrupt these traditional methodological and academic assumptions in the field (Cockburn-Wootten, McIntosh, Smith, & Jefferies, 2018; Ramanayake, McIntosh, & Cockburn-Wootten, 2018; Rydzik, Pritchard, Morgan, & Sedgley, 2013; Scarles, 2010). As critical tourism researchers, we have sought to adopt methods that foster creative approaches, develop tactics for reciprocal knowledge transfer and to encourage ourselves and 'our students to think against the grain' (Singh, 2012, p. 23). Our paper seeks to contribute to this gap by offering a consideration of a research method, Ketso, that we employed to foster creative and reciprocal knowledge transfer between researchers and the stakeholders aiming to make a difference to our communities.

Recently, there has been interest around developing research practices that make an impact and bridge academia, local communities and society. Academics and industry participants, for example, have discussed ways around how to increase research impact to provide benefits to stakeholders outside of academia (Boaz, Fitzpatrick, & Shaw, 2009; Hill, 2016; Reed, 2018). These relationships, illustrate the reality of working within contexts with differing interpretations, sense-making, dynamics, messiness and calls for researchers to consider 'tools and methods that enhance our understandings complex government-business-civil-society relations' (Dredge, 2006, p. 564). Nowotny, Scott, and Gibbons (2003) note that this approach shifts the research focus and dissemination from the 'old paradigm of scientific discovery' to a 'new paradigm of knowledge production' focusing on knowledge distribution, transdisciplinary application-oriented research and considerations of multiple accountabilities (p. 179). This perspective calls for academics to position researchers and participants as facilitators and co-creators, rather than 'lone experts' engaged in top-down and one-way communication (Cockburn-Wootten et al., 2018).

Scholars, for example, Galvagno and Dalli (2014), Greenhalgh, Jackson, Shaw, and Janamian (2016) and Campos, Mendes, Valle, and Scott (2018) have presented systematic reviews of the literature on co-creation. As a concept prevalent in the management, psychology, education, planning and development literature, co-creation focuses on creating value (both materially and symbolically) through interaction for collaborative knowledge generation and the development of new opportunities. Co-creation is a concept increasingly being used by academics as an approach to align research with an understanding and engagement with end users. As a method, it focuses on civic engagement, power sharing, intersectional collaboration, processes, relationships and conflict management. It is seen as an approach to achieve research impact; to move 'beyond the ivory towers' to deliver social impact (Greenhalgh et al., 2016, p. 392). To illustrate this approach, we provide a critical review of the Ketso method for co-creating tourism knowledge. Ketso is a portable toolkit, or 'workshop in a bag', that enables people to think and work together more productively in a facilitated environment. Our twofold critical reflections on Ketso provide an overview of the Ketso toolkit as a method of data collection to co-create solutions for accessible tourism through stakeholder engagement and as a supplementary data analysis tool used in our tourism research.

Moving from traditional methods to co-creative disruption

Methodologies provide a framework for the philosophical approach used within research and shape how the researcher uses data collection tools. It is this theoretical approach that drives the design and handling of the data collection tools. For instance, two researchers can both employ interviews in their studies, but how they design the procedures, questions, relationship with participants, ethical considerations and analysis techniques will all vary depending on their philosophical premise.

Within the wider academia, scholars have critiqued research designs for their 'cookie cutter' approach in which the same designs are repetitively and narrowly applied in ways that limit what is studied and how it is studied' (Harwell, 2011, p. 147). Across disciplines, similar constraints with traditional research tools have been identified. For example, consumer research scholars Zaltman and Coulter (1995) noted that traditional research techniques still limit the experiences and participation of the participant due to the dominance of written and verbal responses; the framework in survey research and in most qualitative techniques is verbocentric (p. 36). Within health, researchers have noted that despite the multitude of research studies on trying to change health outcomes for communities, very little change has actually occurred in the communities facing those health issues (Horowitz, Robinson, & Seifer, 2009). Instead, health researchers have realised that it is crucial to include participants in the research design process to increase self-efficacy and mutual learning around the issue.

Traditional research techniques are also limiting for sustainability and environmental disciplines due to the complex problems encountered within this sector. Sustainability scholars require tools that overcome problems, in an ambiguous context, with an overall desire to vision new solutions and possibilities (Dale & Newman, 2005; Haseman, 2006). Similarly, sustainability and tourism scholars have also argued for a reorientation to include research tools that value social relations, appreciate the diversity of context and enhance active involvement in the study design for research participants (Cockburn-Wootten et al., 2018; Dredge, 2006; Potts & Harrill, 1998). In response to these limitations, scholars have sought to employ and encourage consideration of designs and tools that challenge orthodox enquiry.

Participatory approaches to engage diverse stakeholders have been seen as a more equitable way to conduct research, address the issue or need facing the community, develop mutual learning and gain social change (Cockburn-Wootten et al., 2018). This framework moves away from seeking one truth to trying to understand how issues are 'socially constructed, therefore that it is subject to reinterpretation, revision and enrichment' and is concerned with praxis and transformation (Fals Borda, 2001, p. 31). In seeking this change, researchers are concerned with the theoretical orientation of the research and how to create participation. This framework challenges traditional considerations around the concept of expertise and participation, i.e. who can know and make decisions about the topic under scrutiny in the project (Bean & Baber, 2011). To facilitate this approach, creative participatory methods are employed to stimulate dialogue, develop understandings that transcend verbal and written language descriptors, aiming to create transformation and change solutions, especially for complex issues facing diverse communities.

Creative participatory methods are defined as context-specific and 'challenge dominant assumptions and conventions around what constitutes research, knowledge and impact' (van der Vaart, van Hoven, & Huigen, 2018, p. 6). In challenging dominant assumptions, creative and participatory methods open up new understandings and can reveal emotional responses too. Examples show that the use of creative participatory methods allows dialogue, helps identify the problem or barriers, indicates different interpretations of an issue and deeper understandings of experiences that may be emotional or difficult to describe. Photovoice, for instance, enables participants to discuss experiences that are sensitive, emotional and topics that cannot be captured or labelled. Baker and Wang (2006) study examining chronic pain concluded that this research tool developed self-efficacy to communicate the experience of chronic pain. Another example, LEGO® SERIOUS PLAY®, aims to encourage dialogue and solutions through the use of models built by participants. It heavily draws on concepts of play, the use of metaphors, flow and constructivism (Wengel, McIntosh, & Cockburn-Wootten, 2016).

Creative participatory approaches differ from the mainstream tools because they do not rely on written or verbal responses, foreground accountability, involvement and mutual learning of individuals that experience and are involved in the research issue. These types of studies are not researcher-led or dominated; instead, they are designed, to begin with the participants identifying and prioritising the issues that they see as important to their communities. This approach aims to develop inductive dialogic communication processes to 'create space[s] for the dynamic interchange of knowledge and understandings' through 'a shared commitment to understanding issues and processes … to construct new ways of conceptualising practice' (Cook, Atkin, & Wilcockson, 2017, p. 1). Working with communities and ensuring that dissemination of mutual knowledge and learning outcomes are crucial for this methodological research design.

Within tourism studies, there has been a recent move to design research that considers the role of researchers and participants as co-creators of knowledge outcomes. The premise framing this approach is a desire to adopt methods that decentralise power, foster creative thinking and develop knowledge transfer and dissemination. This framework draws on a variety of critical theoretical approaches within tourism, such as hopeful tourism, feminist, indigenous to social justice approaches to research design (Canosa, Wilson, & Graham, 2017; Hales, Dredge, Higgins-Desbiolles, & Jamal, 2018; Pritchard, Morgan, & Ateljevic, 2011). A table below provides a summary of creative participatory methods used in tourism research (Barry, 2017; Ji & King, 2018; Ren, Pritchard, & Morgan, 2010; Rydzik et al., 2013; Salazar, 2012; Wengel et al., 2016; Willson, McIntosh, & Zahra, 2013) (Table 1).

Participants involved in creative participatory approaches to research report enjoyment during the process highlighting that creative participatory methods allow time to reflect on their personal experiences and greater focus on 'doing' activities, hence, producing richer context than question and answer interviews (Banks, 2007; Pink, 2012). Yet, despite the advantages 'visual research methods … remain[s] reasonably marginal within an existing qualitative practice' with the prevalence of traditional qualitative approaches (Gauntlett & Awah, 2012, p. 590). The majority of this work aims to start with an inductive, emic approach endeavouring to include self-efficacy, dialogue and 'admit the possibility of the existence of other visions of the world where nature, spirituality and human relationships play a leading role in shaping the conformation of knowledge' (Espeso-Molinero, Carlisle, & Pastor-Alfonso, 2016, p. 1334). With this in mind, we present a discussion of Ketso, a creative participatory tool.

Table 1. Creative participatory methods.

Method	Degree of participants participation
Elicitation based on photo methods	Participants produce a photo which represents participants' voice
Drawing based methods	Participants produce a drawing or a map with own drawings
Collage based methods	Participants produce a collage using existing pictures and text
Mind mapping and diagramming methods	Participants produce maps and diagrams
Sounds and music-based methods	Participants required to record sounds and music
Narrative based methods	Participants write down their stories, poems and narratives
Artefact based methods	Participants produce creative artefacts

Source: Authors.

The philosophy and process of Ketso

Ketso is a toolkit which enables people to think and work together more productively (Tippett, 2013; Tippett & How, 2011). The philosophy behind the method embraces principles of participatory research grounded on co-creational practices in order to give (marginalised) participants individual voice to create an impact on the project for the wider community. As a participatory action research tool within the social sciences (Tippett, Handley, & Ravetz, 2007), the technique draws on theories of creative thinking (De Bono, 2009), mind mapping (Buzan & Buzan, 2006), experiential learning (Kolb, 1984), and multiple intelligences (Gardner, 1999). Originating from the disciplines of education and environmental studies, Ketso has now been used in various disciplines to create engagement, co-learning and collaborative thinking (Bates, 2016). In one study on organisational change in academic libraries, for instance, Ketso was used as a tool to develop research questions and for mapping change within libraries spaces (Whitworth, Torras I Calvo, Moss, Amlesom Kifle,

& Blåsternes, 2014). In their research, participants took part in six Ketso workshops adopting mapping techniques to identify changes in their workplace. During the workshops, participants followed the steps of Ketso method to discuss their current working situation, information needed in order to complete work tasks, sources of information and challenges of acquiring this information. At the end of the workshop, participants prioritised future actions which were revisited during later workshops. Data obtained during these Ketso sessions allowed broad comparisons of the organisational changes in the workplace environment. Internationally, Ketso is now used on six continents around the world, and workshop themes have included community-led planning and regeneration, engaging stakeholders on behalf of local and government agencies, corporate training, developing new businesses, team building, student-led learning, and providing tools for teachers and researchers (Tippett & How, 2011).

In tourism studies, Ketso is still relatively new and underexplored. McIntosh and Cockburn-Wootten (2016) have used Ketso as a qualitative data collection tool for engaged, participatory tourism scholarship with a variety of diverse participants to address social issues. They argue that Ketso offers a creative way for tourism researchers to become facilitators in co-creating insightful outcomes with tourism stakeholders and the wider community. Furthermore, Ketso was used to understand the concept of community hospitality and how this type of hospitality is employed by refugee-service organisations in facilitating welcome to refugees to New Zealand (McIntosh & Cockburn-Wootten, 2018). In another study, Cockburn-Wootten et al. (2018) discuss and illustrate how these type of creative tools can provide new possibilities for practice, knowledge and crossing traditional tourism stakeholder silos. They conclude by arguing that orientating research studies within a participatory creative framework by using tools such as Ketso, enables us to challenge dominant assumptions, create reciprocal learning opportunities and make a difference to our communities.

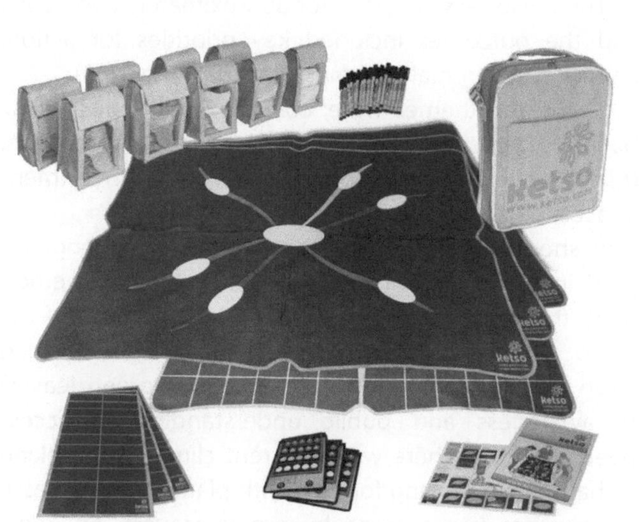

Figure 1. Ketso Kit.
Source: www.ketso.com. Used with permission.

As a facilitated workshop technique, Ketso represents an inclusive tool which helps to unleash participants' creativity. Inclusive research denotes tools and activities that involve people beyond their traditional role in mainstream research as a 'subject'. These activities and roles can encompass a broad range of involvement to include framing research questions, providing the right to access information, leadership roles with the study to involvement in the data analysis (Bigby, Frawley, & Ramcharan, 2014). Ketso enables many of these participatory, inclusive activities to happen by promoting critical dialogue to identify the key issues, possible collaborative solutions and develops a reciprocal knowledge space for individuals, thinking through issues together. To illustrate how these activities occur through Ketso, we begin by describing the toolkit and process of Ketso.

A standard Ketso kit accommodates up to 24 participants, and consists of a large felt mat, grid mat, coloured plastic cards ('leaves') and icons, felt stripes ('branches'), marking pens with water-soluble ink, and the guide (www.ketso.com). The leaves, branches, and icons are movable and attach to felt with Velcro. The kit (Figure 1) is sustainable; all items are reusable, colourful, and tactile. As an accessible, inclusive tool, Ketso could be used with illiterate participants who can make drawings on leaves and with colour-blind participants who can determine the colour of the leaf from the designated letter at the corner of each leaf.

The Ketso workshop is based on a metaphorical analogy of a tree. The analogy provides a universal understanding of the elements of a tree during its growth. Some participants of previous Ketso workshops have agreed that the metaphorical use of a tree was excellent at cultivating a natural flow of discussion during the workshop from initial growth in the soil, to development of branches and leaves (Lombard, 2016). The centrepiece of the workshop is represented by the 'trunk'; the 'branches' represent themes, and the 'leaves' represent different ideas expressed by the participants.

Usually, the Ketso session starts with a warm-up question. The questions guiding the session theme are asked one by one. To answer each question, participants are allocated a specific time frame, for example, 10–15 min. Commonly, each session has four key stages (Figure 2).

Each stage is associated with a different question and 'leaf' colour. To answer the question, participants write/draw the answer onto the 'leaf' and place it on a relevant 'branch' in order to co-create the themes.

At the first stage of the session, participants receive brown leaves, which represent soil on which ideas will grow. Participants could be asked the following questions: What works well? What do people already do that is effective? What is good about what we do now? After some time for generating ideas in silence, participants are asked to share their ideas one after another, one idea at a time.

The next stage is represented by green leaves and aims to generate new and creative ideas. Participants are encouraged to provide an answer to what could be done differently. The metaphor here is green shoots as newly growing ideas. If during the sharing stage participants mention similar ideas, they may be clustered on a branch. The third stage of the session uses grey leaves, which represent challenges and barriers. Thinking of a tree analogy, grey leaves represent clouds hiding the sun and getting in the way of what participants are aiming to achieve. At this stage, participants need to answer the following questions: What are the key barriers and challenges? What gets in the way or makes things difficult? To overcome these challenges, participants are invited to use green leaves to identify particular solutions to those problems. The last stage of the session relates to the coming out of the bright sunshine that drives growth and makes new ideas happen. Here the questions are focused on the goals which participants would like to achieve in the future.

Ketso encourages participants to engage in productive co-creative dialogue. The method allows each participant to contribute to the discussion and engage in the co-creation of the content. One after each other participants present their ideas in a balanced environment where the shy participants are getting their voice heard and dominant participants have limited time to speak out their ideas. When participants see their ideas taking shape, alongside all the thoughts of the other participants, they are encouraged to move the leaves on the felt mat in front of them in order to see the commonalities and/or different points of view and to make new connections (Tippett & How, 2011).

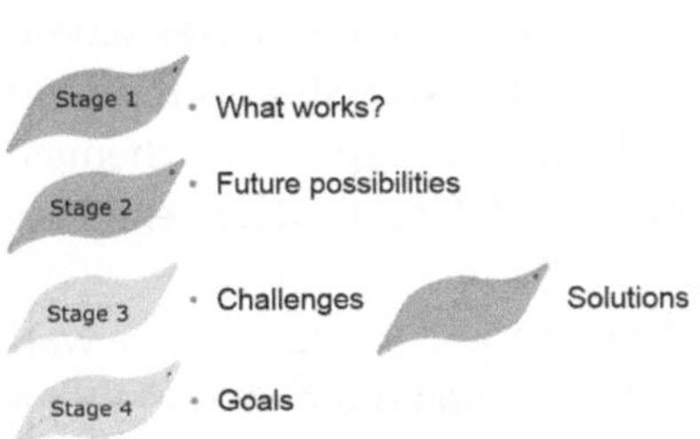

Figure 2. Stages of Ketso Session.
Source: www.ketso.com. Used with permission.

In previous research, Ketso has been used as a tool for data collection and as a tool for PhD students' training and development (Ketso, 2018; Njiraini, 2015). In this paper, we critically review Ketso as a co-creative community collaborative tool with stakeholders in the New Zealand *Tourism For All* project. Furthermore, we extend the use of the Ketso method and propose Ketso as a complementary data analysis tool.

Ketso as data collection tool: The New Zealand *Tourism For All* project

With the aim to creatively and collaboratively engage stakeholders to consider future solutions for achieving travel that is accessible for people with disabilities, the second and third authors facilitated a round-the-table planning session using Ketso with stakeholders from the tourism and access sectors in Christchurch in the South Island of New Zealand; a city being rebuilt following a series of devastating earthquakes. The aim of the Ketso session was to engage the stakeholders in a meaningful collaborative thinking process to consider and plan priorities and actions for accessible tourism development in the city and its wider region. Participants included four representatives from access organisations, one City Council representative, two representatives from major local tourist attractions, and one individual resident interested in the topic. The participants included those individuals with a disability and those without, but this was not a focus for their participation, and the question of their own disability was not raised with participants unless they themselves raised it.

The Ketso session lasted for approximately two hours, and the outcomes included key priorities for action, involving seven main themes/branches (Figure 3).

The seven themes were: communication; awareness but not yet understanding; innovation; celebration; quality; action; and promoting the return on investment in accessible tourism.

It should be noted that the Ketso method engages participants in co-creating and naming their own emergent 'themes' from the process. Overall, findings of the Ketso session revealed that participants felt there had been a positive growth (represented by green leaves) in awareness and public understanding of access needs and that there was a current climate for making a change and moving forward with planning for accessible tourism (represented by brown leaves). An unexpected outcome of the session was that few barriers were reported (represented by grey leaves) to achieving

Figure 3. Ketso session: The New Zealand *Tourism For All* project.
Source: Authors' Image.

an accessible tourism plan for Christchurch, other than the low participation rate of stakeholders in the discussion about planning for accessibility. One of the objectives of the Ketso method is not to overly focus on barriers but to instead empower an innovative process for creative, collaborative thinking, and this may be a reason for the greater focus given to creative thoughts for future solutions, potentially fuelled by a shared positive attitude toward social change among the stakeholders who participated in the session. The key priorities for the future (represented by the yellow leaves) were the need for advertising campaigns that include people with disabilities; knowledge sharing for greater awareness of access needs; generation of pride from leading by example to make a difference (modelling excellent access design and 'telling the world'); and demonstrating and showcasing success in accessible tourism. Given the priorities identified by the participants to promote access champions and raise greater awareness of access needs, with the consent of participants, a final documentary, 'Take the Time', was developed and shared on social media and in various public venues to showcase the outcomes of the project, including footage of the Ketso session as the heart of the co-created action plan (see: https://www.youtube.com/watch?v=xux0nKY2MQ0&t=2s).

Ketso as a supplementary analysis tool

In this section, we would like to provide a critical overview on the combination of thematic analysis with the Ketso method to co-create the research findings of our tourism research (Wengel, McIntosh, & Cockburn-Wootten, 2018). Thematic analysis is a flexible approach which can be applied across a range of epistemologies and research questions (Braun & Clarke, 2006). It is an established tool for qualitative analysis which is based on searching through the collected data in order to find frequent patterns. The analysis is versatile and moves 'beyond counting explicit words or phrases and focus[es] on identifying and describing both implicit and explicit ideas within the data, that is, themes'

(Guest, MacQueen, & Namey, 2012, p. 10). King (2004) argues that an advantage of thematic analysis is the opportunity to examine the perspectives of research participants, emphasising the similarities and differences, and generating unanticipated insights. Despite the advantages of thematic analysis, it is important to acknowledge its disadvantages.

Some researchers critique thematic analysis by arguing that novice researchers may feel unsure of how to conduct a rigorous thematic analysis (Nowell, Norris, White, & Moules, 2017). Holloway and Todres (2003) claim that themes derived from the research data, could lack coherence and be inconsistent. As in qualitative research, a researcher is an instrument of analysis who is collecting the data, analysing the data and making interpretation of the data regardless of methods used (Starks & Trinidad, 2007), Nowell et al. (2017) call for more sophisticated tools and transparency in the process to conduct trustworthy thematic analysis.

To strengthen the analysis methods and overcome the limitations of the thematic analysis mentioned above we experimented with Ketso as a supplementary analysis tool in analysing data collected via unstructured in-depth interviews. The data originated from our multi-method qualitative study focusing on the relationships between farmers and volunteers working on their farms (as a part of World Wide Opportunities on Organic Farms (WWOOF) programme). We used an inductive approach to thematic analysis to interpret and elucidate meanings created by the participants. Initially, codes generated by thematic analysis were grouped into 12 common areas. Thereafter we experimented with Ketso as a supplementary analysis tool.

The first author conducted three one-on-one Ketso sessions to analyse the data. During the first session, she focused on codes relating to the relationships between farmers and their volunteers. As the result of the session, she rearranged the themes, and 12 common themes were confirmed. During the second Ketso session, the standard steps of Ketso were applied. The second session focused on identifying and confirming existing common areas as well as synthesising and refining the themes. As a result of the session, eight common themes emerged. After the third session, four key themes were determined and confirmed (Figure 4).

As an outcome of the three Ketso sessions, the data generated by thematic analysis were rearranged into the common areas, and four final themes emerged: dirt, crossing the threshold, ideals, and ethics. Overall, the Ketso sessions helped us to achieve clarity within the common areas produced by thematic analysis. Furthermore, with Ketso as a supplementary analysis tool we gained more credibility in the analysis process and focused our attention on the key aspects of farmers and volunteers relationships and to define the four final themes.

Discussion

In the New Zealand *Tourism For All* project, Ketso was a tool used to creatively and collaboratively engage stakeholders from a range of sectors who might not otherwise share dialogue, such as the tourism and access sectors, in order to co-create an action plan for accessible tourism. As such, the process was crucial for enabling problem-solving around the complex, unconsidered and polysemous issues (Mitra & Buzzanell, 2014). Participants reported gaining awareness and learning as well as being able to share knowledge as a result of being involved. On a wider level, the Ketso process fostered communication and relationship building and facilitated an opportunity to question existing practices. Despite reviewing different forms of stakeholder participation, including public hearings, advisory committees, surveys, focus groups, collaboration, work groups, among others, Byrd (2007) called for future studies to further consider the different forms and methods to achieve greater success in engaging stakeholders in the tourism planning process. Most previous tourism stakeholder analysis has been interview based (e.g. Aas, Paster, Stokes, Olsen, and Dewhirst (2005); Ernawati and Sugiarti (2005); Gillovic and McIntosh (2015)). Alternatively, this project sought a more collaborative approach in recognition of the fact that different forms of stakeholder participation do not all involve the same level of participation. Our objective was to actively engage stakeholders in an innovative process so that they feel empowered to think creatively about future solutions and inclusive to influence the decision. As Byrd (2007, p. 9) pointed out, 'even if parties cannot resolve a particular issue, the process should be able to help them understand the goals and perspectives of others by fostering communication and build relationships'.

Alternative planning tools such as community mapping and focus groups can be seen as limiting because they are said to impede mutual learning between participants; do not allow participants to engage in an active or meaningful manner; focus too much on problems and barriers hence reducing creative thinking; do not focus on future planning or creative thinking for new solutions; do not innovate or animate the process, nor plan for consensus as a mechanism to achieve development outcomes (McIntosh & Cockburn-Wootten, 2016). As a method, Ketso does, however, require all participants to be available at the same time

Figure 4. Final Ketso Session: Four Themes.
Source: Authors' Image.

and location, and this might be considered a major limitation for stakeholder collaboration within a complex and diverse system such as tourism. We found this in our project, when only eight stakeholder participants were able to attend, despite over 200 invitations being sent out.

Previous research has also readily identified the problems of power, and issues of the 'silent majority' or 'most vocal' voice in tourism stakeholder collaborations (Aas et al., 2005; Tosun, 2000). The Ketso tool allowed all members of the session to contribute equally, rather than allowing certain voices to dominate the group; it is founded on inclusive and learning philosophies and systems thinking (see Tippett et al., 2007). A major advantage of the tool is, therefore, its ability to facilitate both individual ideas and group analysis toward a visual plan proposing a solution/action(s). And because Ketso parts are moveable (attached with Velcro), ideas can flow and be developed as they are raised and discussed among the participants. The main themes emerging from the workshop are inductively co-created by the participants themselves throughout the session. The final visual output helps participants understand complex relations; these are easily identifiable from the colour of the leaves in the output for example, where the most good is currently happening (i.e. most brown leaves) and where there is the best potential for growth (i.e. most green and yellow leaves). In short, the tool highlights the process of communication between different stakeholders to value their contribution, identifying the overall themes emerging from the group's thinking and their priority actions.

The attention to the inductive process of communication and knowledge sharing between different stakeholders maximises the value of their contribution and commitment to priority actions is a key benefit of the Ketso toolkit. Indeed, for this reason, McIntosh and Cockburn-Wootten (2016) have advocated it as a useful tool for enabling stakeholders to co-create methods for addressing social issues, such as accessibility and inclusion and proposing actions and solutions. Indeed, since disability is generally regarded as socially constructed (Oliver, 2013), this project highlighted that there remains an important need to reduce social barriers, and especially those barriers currently imposed by the tourism industry. To achieve this, it is argued that accessibility needs to become a fundamental part of the value chain among stakeholders, such as that proposed in the concept of universal design (Buhalis & Darcy, 2010). As a tool, Ketso helps stakeholders to creatively understand the absences in their organisational sector and starts to get them ready for transformation. Ketso discussions make the absences and assumptions in the everyday worldviews of powerful stakeholders evident and open up an agency for changes to the tourism system. Significantly, it is the ontological framing behind the tool that allows stakeholders and academics to identify silences 'to help us see openings, to provide a space of freedom and possibility' (Gibson-Graham, 2008, p. 619). Through providing space for creative thinking around accessibility and tourism, the Ketso process ensured that solutions shaped the kind of environment and tourism spaces that 'we can imagine and create, ones in which we enact and construct' (Gibson-Graham, 2008, p. 619). These collaborative discussions allow the choice to think, act and consider different ways without the cynicism of critique or the restrictions of the past. By gathering together in one place, the stakeholders share tacit knowledge, overcome initial barriers from powerful stakeholders and gain solutions from people with direct experience of the issue.

In our second use of the method, we applied Ketso to ensure greater trustworthiness in the data analysis process; we decided to combine two analysis methods which mutually benefit each other (Guba, 1981), such as thematic analysis and Ketso. Two main benefits of thematic analysis are that it provides a systematic element to data analysis and provides an opportunity to understand the potential of any issue more widely (Marks & Yardley, 2004). Although thematic analysis moves beyond counting words and focuses on identifying and describing implicit and explicit ideas (Namey, Guest, Thairu, & Johnson, 2008), it could be hidden, subjective, and unable to address the complications of reflective research.

These limitations of thematic analysis could, however, be filled by further revisiting the data with Ketso methodology. Ketso enabled us to focus the thinking (Tippett & How, 2011) and to see the deeper connections between different codes as common areas were co-created by the researcher. Using this tool allowed us to merge the common patterns into four key themes by following the logical path outlined in the Ketso methodology. With the help of this co-creative visual tool, we were able to get a deeper understanding of themes, to draw connections, to prioritise themes, and to find patterns which were previously overlooked. Ketso sessions helped us to achieve clarification and to define four final themes by focusing the attention on the key aspects and allowing the researcher to co-create the themes.

Despite the advantages of the combination of Ketso and thematic analysis, there still remain limitations. The main limitations of the Ketso method is the necessity of a special kit, skilled facilitator as well as the space and time for running the session. In previous studies, some research participants found Ketso to be a difficult and unsettling concept because of its inductive nature (McIntosh & Cockburn-Wootten, 2016). For the first author, the difficulty was to facilitate a one-on-one session for herself as a lot of reflexive thoughts intervened on the Ketso process. Such as, 'Am I a facilitator now?' or 'What would be the next step of the Ketso process?' To overcome this difficulty, we would recommend that the Ketso session is facilitated by another person. In this way, a researcher would be guided through the Ketso session by the facilitator and therefore will be able to better concentrate on the context of the session, and focus on synthesising the themes.

Conclusions

Effective social impact that makes a difference to our societies requires a change in research design, theoretical orientation and activities that increase participation and engagement. In this article, we presented Ketso as a complementary qualitative tool which could enhance this type of research design, create opportunities for richer understandings, new knowledge across diverse stakeholders and provide reciprocal learning opportunities. Aligned with Ind and Coates's (2013) understanding of the co-creation concept, Ketso offers researchers, participants and wider stakeholders the space, time and process to generate solutions, build capacity, and create value together. Furthermore, we contribute by providing guidance toward using Ketso to complement other qualitative data analysis methods. For example, Braun and Clarke (2006) established thematic analysis as a foundational method for qualitative data analysis, and although it has been widely used across social sciences, many researchers criticise it for the simplicity and its lack of coherence. Our research findings suggest that a combination of Ketso and thematic analysis can potentially achieve richer saturation across the data which increases the accountability and verification of the research findings and analysis. In our experience combining Ketso and thematic analysis enables a transactive process of data analysis to occur. This provides opportunities for scholars, especially those investigating new topics or issues that participants may have difficulty in expressing their experiences, for both explicit and tacit knowledge to be highlighted.

For tourism researchers dealing with a diverse system involving organisations, NGOs, activists and communities, Ketso contributes by offering opportunities for involving both individual and group reciprocal learning and engagement. Ketso also contributes to the wider academic field by providing a collaborative tool to enable communities to come together, overcome silos and think – enabling them to identify problems and then move towards creating innovative solutions to addresses issues they have identified together. As a tool, Ketso illustrates a co-creative transactive method that can be used by tourism researchers who are theoretically aligned to work with and within communities to co-create change and overcome traditional academic top-down approaches to research. Ketso provides opportunities for researchers to overcome these silos and work differently, to share experiences, problems, expertise, all in order to tackle particular issues identified by the community – not the academic. In doing so, as illustrated in the New Zealand *Tourism For All* research, Ketso facilitates individuals and groups' thinking around identifying problems rather than concentrating on symptoms to then identify solutions.

As a co-creative method, Ketso fits into a range of collaborative, action-oriented transformative research and engagement philosophies that encourage individuals to become involved in thinking about innovative solutions to tackle issues and create positive social impact. These alternative narratives and actions that inductively emerge during the Ketso process help create an understanding for groups of the choices available for tackling the issue. This collaborative framework and the Ketso tool provide value-creation to all involved in the research – not just the more powerful or dominate individuals in the project.

Disclosure statement

No potential conflict of interest was reported by the authors.

ORCID

Yana Wengel http://orcid.org/0000-0002-8131-4137
Alison McIntosh http://orcid.org/0000-0003-1593-700X
Cheryl Cockburn-Wootten http://orcid.org/0000-0002-3339-630X

References

Aas, J. A., Paster, B. J., Stokes, L. N., Olsen, I., & Dewhirst, F. E. (2005). Defining the normal bacterial flora of the oral cavity. *Journal of Clinical Microbiology*, *43*(11), 5721–5732. doi:10.1128/jcm.43.11.5721-5732.2005

Baker, T. A., & Wang, C. C. (2006). Photovoice: Use of a participatory action research method to explore the chronic pain experience in older adults. *Qualitative Health Research*, *16* (10), 1405–1413. doi:10.1177/1049732306294118

Banks, M. (2007). *Using visual data in qualitative research*. Los Angeles, CA: SAGE.

Barry, K. (2017). Diagramming: A creative methodology for tourist studies. *Tourist Studies*, *17*(3), 328–346. doi:10.1177/1468797616680852

Bates, J. S. (2016). Ketso: A new tool for extension professionals. *Journal of Extension*, *54*(1), 1–4.

Bean, G., & Baber, K. M. (2011). Connect: An effective community-based youth suicide prevention program. *Suicide and Life-Threatening Behavior*, *41*(1), 87–97. doi:10.1111/j.1943-278X.2010.00006.x

Bigby, C., Frawley, P., & Ramcharan, P. (2014). Conceptualizing inclusive research with people with intellectual disability. *Journal of Applied Research in Intellectual Disabilities*, *27*(1), 3–12. doi:10.1111/jar.12083

Boaz, A., Fitzpatrick, S., & Shaw, B. (2009). Assessing the impact of research on policy: A literature review. *Science and Public Policy*, *36*(4), 255–270. doi:10.3152/030234209X436545

Braun, V., & Clarke, V. (2006). Using thematic analysis in psychology. *Qualitative Research in Psychology*, *3*(2), 77–101. doi:10.1191/1478088706qp063oa

Buhalis, P. D., & Darcy, S. (2010). *Accessible tourism: Concepts and issues*. Bristol: Channel View Publications.

Buzan, T., & Buzan, B. (2006). *The mind map book*. Harlow, UK: BBC Active.

Byrd, E. T. (2007). Stakeholders in sustainable tourism development and their roles: Applying stakeholder theory to sustainable tourism development. *Tourism Review*, *62*(2), 6–13. doi:10.1108/16605370780000309

Campos, A. C., Mendes, J., Valle, P. O. D., & Scott, N. (2018). Co-creation of tourist experiences: A literature review. *Current Issues in Tourism*, *21*(4), 369–400. doi:10.1080/13683500.2015.1081158

Canosa, A., Wilson, E., & Graham, A. (2017). Empowering young people through participatory film: A postmethodological approach. *Current Issues in Tourism*, *20*(8), 894–907. doi:10.1080/13683500.2016.1179270

Cockburn-Wootten, C., McIntosh, A. J., Smith, K., & Jefferies, S. (2018). Communicating across tourism silos for inclusive sustainable partnerships. *Journal of Sustainable Tourism*, 1–16. doi:10.1080/09669582.2018.1476519

Cook, T., Atkin, H., & Wilcockson, J. (2017). Participatory research into inclusive practice: Improving services for people with long term neurological conditions. *FQS*, *19*(1), 1–26.

Dale, A., & Newman, L. (2005). Sustainable development, education and literacy. *International Journal of Sustainability in Higher Education*, *6*(4), 351–362. doi:10.1108/14676370510623847

De Bono, E. (2009). *De Bono's thinking course*. Harlow, UK: BBC Active.

Dredge, D. (2006). Networks, conflict and collaborative communities. *Journal of Sustainable Tourism*, *14*(6), 562–581. doi:10.2167/jost567.0

Ernawati, D. B., & Sugiarti, R. (2005). Developing an accessible tourist destination model for people with disability in Indonesia. *Tourism Recreation Research*, *30*(3), 103–106. doi:10.1080/02508281.2005.11081492

Espeso-Molinero, P., Carlisle, S., & Pastor-Alfonso, M. J. (2016). Knowledge dialogue through indigenous tourism product design: A collaborative research process with the Lacandon of Chiapas, Mexico. *Journal of Sustainable Tourism*, *24*(8–9), 1331–1349. doi:10.1080/09669582.2016.1193188

Fals Borda, O. (2001). Partipatory (action) research in social theory: Orgins and challenges. In P. Reason, & H. Bradbury (Eds.), *Handbook of action research participative inquiry & practice* (pp. 27–47). London, UK: Sage.

Galvagno, M., & Dalli, D. (2014). Theory of value co-creation: A systematic literature review. *Managing Service Quality: An International Journal*, *24*(6), 643–683. doi:10.1108/MSQ-09-2013-0187

Gardner, H. (1999). *Intelligence reframed: Multiple intelligences for the 21st century*. New York, NY: Basic Books.

Gauntlett, D., & Awah, F. (2012). Action-based visual and creative methods in social research. In I. Heywood, & B. Sandwell (Eds.), *The handbook of visual culture* (pp. 589–605). Oxford, UK: Berg Publishers.

Gibson-Graham, J. K. (2008). Diverse economies: Performative practices for 'other worlds'. *Progress in Human Geography*, *32*(5), 613–632. doi:10.1177/0309132508090821

Gillovic, B., & McIntosh, A. (2015). Stakeholder perspectives of the future of accessible tourism in New Zealand. *Journal of Tourism Futures*, *1*(3), 223–239. doi:10.1108/JTF-04-2015-0013

Gillovic, B., McIntosh, A., Darcy, S., & Cockburn-Wootten, C. (2018). Enabling the language of accessible tourism. *Journal of Sustainable Tourism*, *26*(4), 615–630. doi:10.1080/09669582.2017.1377209

Greenhalgh, T., Jackson, C., Shaw, S., & Janamian, T. (2016). Achieving research impact through co-creation in community-based health services: Literature review and case study. *The Milbank Quarterly*, *94*(2), 392–429. doi:10.1111/1468-0009.12197

Guba, E. G. (1981). Criteria for assessing the trustworthiness of naturalistic inquiries. *Educational Communication and Technology*, *29*(2), 75–91. doi:10.1007/BF02766777

Guest, G., MacQueen, K. M., & Namey, E. E. (2012). *Applied thematic analysis*. Los Angeles, CA: SAGE.

Hales, R., Dredge, D., Higgins-Desbiolles, F., & Jamal, T. (2018). Academic activism in tourism studies: Critical narratives from four researchers. *Tourism Analysis*, *23*(2), 189–199. doi:10.3727/108354218X15210313504544

Harwell, M. (2011). Research design in qualitative/quantitative/mixed methods. In C. F. Conrad, & R. C. Serlin (Eds.), *The Sage handbook for research in education: Pursuing ideas as the keystone of exemplary inquiry* (pp. 147–164). Thousand Oaks, CA: Sage.

Haseman, B. (2006). A manifesto for performative research. *Media International Australia Incorporating Culture and Policy*, *118*(1), 98–106. doi:10.1177/1329878X0611800113

Hill, S. (2016). Assessing (for) impact: Future assessment of the societal impact of research. *Palgrave Communications*, *2*, 16073. doi:10.1057/palcomms.2016.73

Holloway, I., & Todres, L. (2003). The status of method: Flexibility, consistency and coherence. *Qualitative Research*, *3*(3), 345–357. http://dx.doi.org/10.1177/1468794103033004

Horowitz, C. R., Robinson, M., & Seifer, S. (2009). Community-based participatory research from the margin to the mainstream: Are researchers prepared? *Circulation*, *119*, 19.

Ind, N., & Coates, N. (2013). The meanings of co-creation. *European Business Review*, *25*(1), 86–95.

Ji, M., & King, B. (2018). Explaining the embodied hospitality experience with ZMET. *International Journal of Contemporary Hospitality Management*, *30*(11), 3442–3461. doi:10.1108/IJCHM-10-2017-0709

Ketso. (2018). *Research- Ketso as a method*. Retrieved from https://ketso.com/examples-case-studies/case-studies/research#DTC

King, N. (2004). Using templates in the thematic analysis of text. In C. Cassell, & G. Symon (Eds.), *Essential guide to qualitative methods in organizational research* (pp. 257–270). London: Sage.

Kolb, D. A. (1984). *Experiential learning: Experience as the source of learning and development*. Englewood Cliffs, NJ: Prentice-Hall.

Lombard, M. (2016). *Rights to the City*. Retrieved from http://www.ketso.com/examples-case-studies/case-studies/research

Marks, D., & Yardley, L. (2004). *Research methods for clinical and health psychology*. Thousand Oaks, CA: SAGE.

McIntosh, A., & Cockburn-Wootten, C. (2018). Refugee-focused service providers: Improving the welcome in New Zealand. *The Service Industries Journal*, 1–16. doi:10.1080/02642069.2018.1472243

McIntosh, A. J., & Cockburn-Wootten, C. (2016). Using Ketso for engaged tourism scholarship. *Annals of Tourism Research*, *56*, 148–151. doi:10.1016/j.annals.2015.11.003

Mitra, R., & Buzzanell, P. M. (2014). Introduction: Organizing/communicating sustainably. *Management Communication Quarterly*, *29*(1), 130–134. doi:10.1177/0893318914563573

Namey, E., Guest, G., Thairu, L., & Johnson, L. (2008). Data reduction techniques for large qualitative data sets. In G. Guest, & K. M. MacQueen (Eds.), *Handbook for team-based qualitative research* (pp. 137–160). Lanham, MD: Altamira.

Njiraini, N. N. K. (2015). *Exploring the importance of critical thinking in creating capabilities for self-reliance in international community development: A Kenyan context*. Glasgow: University of Glasgow.

Nowell, L. S., Norris, J. M., White, D. E., & Moules, N. J. (2017). Thematic analysis: Striving to meet the trustworthiness criteria. *International Journal of Qualitative Methods*, *16*(1), 1609406917733847. doi:10.1177/1609406917733847

Nowotny, H., Scott, P., & Gibbons, M. (2003). Introduction: 'Mode 2' revisited: The new production of knowledge. *Minerva, 41* (3), 179–194. doi:10.1023/a:1025505528250

Oliver, M. (2013). The social model of disability: Thirty years on. *Disability & Society, 28*(7), 1024–1026. doi:10.1080/09687599.2013.818773

Pink, S. (2012). *Advances in visual methodology*. Los Angeles: SAGE Publications.

Potts, T. D., & Harrill, R. (1998). Enhancing communities for sustainability: A travel ecology approach. *Tourism Analysis, 3*(3–4), 133–142.

Pritchard, A., Morgan, N., & Ateljevic, I. (2011). Hopeful tourism: A new transformative perspective. *Annals of Tourism Research, 38*(3), 941–963. doi:10.1016/j.annals.2011.01.004

Ramanayake, U., McIntosh, A., & Cockburn-Wootten, C. (2018). Loss and travel: A review of literature and current understandings. *Anatolia, 29*(1), 74–83. doi:10.1080/13032917.2017.1387162

Reed, M. S. (2018). *The research impact handbook* (2nd ed.). Huntly: Aberdeenshire.

Ren, C., Pritchard, A., & Morgan, N. (2010). Constructing tourism research: A critical Inquiry. *Annals of Tourism Research, 37*(4), 885–904. doi:10.1016/j.annals.2009.11.006

Rydzik, A., Pritchard, A., Morgan, N., & Sedgley, D. (2013). The potential of arts-based transformative research. *Annals of Tourism Research, 40*, 283–305. doi:10.1016/j.annals.2012.09.006

Salazar, N. B. (2012). Tourism imaginaries: A conceptual approach. *Annals of Tourism Research, 39*(2), 863–882. doi:10.1016/j.annals.2011.10.004

Scarles, C. (2010). Where words fail, visuals ignite: Opportunities for visual Autoethnography in tourism research. *Annals of Tourism Research, 37*(4), 905–926. doi:10.1016/j.annals.2010.02.001

Sedgley, D., Pritchard, A., & Morgan, N. (2011). Tourism and ageing: A transformative research agenda. *Annals of Tourism Research, 38*(2), 422–436. doi:10.1016/j.annals.2010.09.002

Singh, T. V. (2012). *Critical debates in tourism*. Bristol, UK: Channel View Publications.

Starks, H., & Trinidad, S. B. (2007). Choose your method: A comparison of phenomenology, discourse analysis, and grounded theory. *Qualitative Health Research, 17*(10), 1372–1380.

Tippett, J. (2013). Creativity and learning – participatory planning and the co-production of local knowledge. *Town & Country Planning, 82*(10), 439–442.

Tippett, J., Handley, J. F., & Ravetz, J. (2007). Meeting the challenges of sustainable development—A conceptual appraisal of a new methodology for participatory ecological planning. *Progress in Planning, 67*(1), 9–98. doi:10.1016/j.progress.2006.12.004

Tippett, J., & How, F. (2011). *Ketso guide*. Manchester, UK: Ketso.

Tosun, C. (2000). Limits to community participation in the tourism development process in developing countries. *Tourism Management, 21*(6), 613–633. doi:10.1016/S0261-5177(00)00009-1

van der Vaart, G., van Hoven, B., & Huigen, P. P. P. (2018). Creative and arts-based research methods in academic research: Lessons From a Participatory Research Project in the Netherlands. *FQS, 19*(2), 1–30. doi:10.17169/fqs-19.2.2961

Wengel, Y., McIntosh, A., & Cockburn-Wootten, C. (2018). Tourism and 'dirt': A case study of WWOOF farms in New Zealand. *Journal of Hospitality and Tourism Management, 35*, 46–55. doi:10.1016/j.jhtm.2018.03.001

Wengel, Y., McIntosh, A. J., & Cockburn-Wootten, C. (2016). Constructing tourism realities through LEGO® SERIOUS PLAY®. *Annals of Tourism Research, 56*, 161–163. doi:10.1016/j.annals.2015.11.012

Whitworth, A., Torras I Calvo, M. C., Moss, B., Amlesom Kifle, N., & Blåsternes, T. (2014). Changing libraries: Facilitating self-reflection and action research on organizational change in academic libraries. *New Review of Academic Librarianship, 20*(2), 251–274. doi:10.1080/13614533.2014.912989

Willson, G. B., McIntosh, A. J., & Zahra, A. L. (2013). Tourism and spirituality: A phenomenological analysis. *Annals of Tourism Research, 42*, 150–168. doi:10.1016/j.annals.2013.01.016

Zaltman, G., & Coulter, R. H. (1995). Seeing the voice of the customer: Metaphor-based advertising research. *Journal of Advertising Research, 35*(4), 35–51.

Co-creating an integrated curriculum alongside community partners: a creative analytic approach

Karla Boluk, Meghan Muldoon and Corey Johnson

ABSTRACT

Traditional approaches to university education are not effective. Problematically, the emphasis in education is *teaching* rather than the ability for students to *learn*. Recognising the constraints of traditional instruction, we assembled a collegial team exploring a creative pedagogical tool to co-instruct and co-learn from our practice. Specifically, our contribution explores our process of co-creating an Integrated Curriculum Design (ICD) project together with our community partners. Our ICD reflects our response to a concern brought forward by a local Destination Marketing Organisation, and was mutually informed by professors, graduate students, and adapted based on the feedback of our undergraduate students, and iterative consultation with our community partners. We employ Creative Analytic Practice (CAP) as a series of vignettes to aid in our understanding of how we co-created our ICD model and the value it has generated for all those involved, pausing at a scholarly-community intersection to reflect on opportunities in co-creating tourism pedagogy with an array of scholarly and community partners.

Introduction

This paper is a continuation of work examining the challenges and benefits of an integrated curriculum design (ICD) from the undergraduate student perspective, and a teaching team representing the perspective of a junior scholar, established scholar, and seasoned scholar (close to retirement) (Boluk, Muldoon, & Johnson, 2019). Building on this work, the aim of this paper is to investigate the process of co-designing and co-constructing an ICD alongside our community partners. Specifically, we reflect on establishing a relationship with our partners, using industry problems to frame assignments drawing on student capabilities, and consciously reflecting on our pedagogical practice to sustain, and build on, our relationships with community partners, while ensuring that student products are meaningful for both groups.

In the winter term of 2016, we carried out an initial exploration of possible courses to integrate in our Tourism Development Programme in an effort to make our curriculum and assessments more relevant, engaging, and consciously focus on the skills development of our students. In this context, our use of skills development refers to building on pre-existing skills acquired by students in their previous courses or tourism experiences outside of class. For example, first-year students will have likely experienced recreation and tourism in some capacity, and learned some basic theory of the field. A next step may encourage students to apply their theoretical knowledge and explain to a peer what could improve their experience. By reflecting and then engaging in dialogue with their peers they are able to think about their position and practice the language used in the field, gaining confidence in their ability to communicate. From here they may critique their experience as a whole. Critically articulating their experiences may draw attention to ways to improve recreation and tourism experiences which could mobilise critical praxis when students are employed in the field. As such, in order to build on students' pre-existing skills we were interested in providing unique experiences aimed at challenging their thinking and growing their skill sets. Increasingly, today's employers are looking to hire people who are flexible, creative, solution-oriented, and able to work in a variety of settings, skills which ICDs are uniquely positioned to help students foster (Athavale, Davis, & Myring, 2008; Huber, Hutchings, & Gale, 2005; Walker & Black, 2000).

Our investigation resulted in identifying a 1st year *Introduction to Tourism* course comprised of 350 students and a 4th year *Advanced Seminar in Tourism, Well-being, and Community* of 22 students. In the past, both courses were scheduled as individual, stand-alone courses. However, they are now framed by ICD, partly in response to a problem brought forward by Regional Tourism Organisation 4 (RTO4), a Destination Marketing Organisation, and now our industry partner. The problem framed by RTO4 was: How may we improve the user experience of millennial consumers? (Boluk et al., 2019). This query now frames our ICD and the common capstone project called Ontario Attractions, Wonders, and Experiences (AWE) between the two cohorts of students who we have positioned as *expert users* (1st years) and *expert designers* (4th years). The curriculum we developed and implemented reflected several of the points indicated on Powell et al.'s (2013) Continuum of Collaboration within a Faculty Discipline such as cross-linking assignments, blending individual courses, and unifying a cohort with a teaching team.

Our ICD framed a multi-phase project with multiple points of intersection between the two tourism cohorts including:

1. A joint guest lecture from our industry partner, RTO4;
2. The development of a UX survey (by our 4th years for our 1st years to respond to) and on-site visits (by our 1st years) to one of five attractions (chosen by our 4th years) and evaluation of their UX satisfaction pre-, during, and post-visit, recording responses on the surveys;
3. Following the receipt of the data generated by the surveys, 4th year students reflected on their initial understandings and assumptions and the two cohorts were then brought together to allow the 4th year students to check the accuracy of their assumptions and probe 1st year students further, as well as brainstorm ideas for potential recommendations.
4. AWE culminated in formal presentations delivered by the 4th year students to the 1st year students, as well as our industry and community partners.

Following an analysis of the UX surveys our expert designers (4th years) offered recommendations to improve the millennial UX. The five featured attractions were invited to participate on a panel during these presentations in-class to query any of the ideas presented by our students and follow up on any ideas that required explanation. Three of our five community partners (attractions) attended the 90-minute presentations along with our industry partner, RTO4. Following the student presentations, we organised a networking lunch for our 4th-year students, industry partner, community partners, and graduate teaching assistants to mutually carry on the dialogue about improving millennial UX, and provide further insights to action change at the participating attractions.

Background: curricular innovations

Traditional university approaches to teaching are defined by their structure (Fenton & Gallant, 2016), one which is immediately familiar to anyone with a passing awareness of post-secondary education. Characterised by Freire (1970) as the banking model of education, instructors present content on a particular topic, generally independently of any other courses students may be taking, and students are expected to learn this information (Fenton & Gallant, 2016). Students are conceived of as passive recipients of expert knowledge (Dewey, 1986) and are rarely given the opportunity to apply the material in real-world contexts or reflect critically upon its meanings (Anderson, 2013; Fenton & Gallant, 2016). Baldwin, Mainieri, and Brookover (2013) likened this approach to higher education as 'a prison to innovation and refuge to silos' (p. 65).

The experiment discussed in this paper, and outlined in greater detail below, is one of integrated curriculum design (ICD). The purpose of an ICD is to break down stagnant silos, collapse barriers between disciplines and subjects to facilitate greater applicability in the material that students are engaging with, and appreciate the relevance of course material to the real world (Fenton & Gallant, 2016; Shoemaker, 1989). Also referred to as integrative learning, this approach seeks to integrate learning experiences across the high school/post-second divide, between cohorts, through introductory to advanced courses, and within the classroom, as well as in real-world spheres (Anderson, 2013; Klein, 2005; Malik & Malik, 2011). As such, an ICD abolishes the banking model style of education whereby the instructor is positioned as the expert and students as empty vessels waiting to be filled with knowledge (Ackoff & Greenberg, 2008; Freire, 1970). An ICD supports students in synthesising information across disciplines, applying theory to practice, developing transdisciplinary problem-solving skills, incorporating multiple perspectives and points of view, and understanding the pathways that their learnings are taking across the entire curriculum (Anderson, 2013; Huber et al., 2005; Miller, 2005; Shoemaker, 1989).

Adopting an integrated approach to designing and delivering curriculum offers a number of benefits. Students engaged in these types of problem-based, team, and experiential learning environments are better able to 'ask meaningful questions about complex problems, locate multiple sources of knowledge, and to compare and contrast knowledge, information and perspectives' (Anderson, 2013, p. 7). Further, students are found to have developed 'improved interpersonal skills, increased engagement, and improved academic learning performance' (Anderson, 2013, p. 7). Integrated curriculum design 'stimulates deep rather than superficial learning' (Malik & Malik, 2011, p. 99) and encourages students to learn at higher levels of thought (VanTassel Baska & Wood, 2010, p. 345). Increasingly, today's employers are looking to hire people who are flexible, creative, solution-oriented, and able to work in a variety of settings, skills which ICDs are uniquely positioned to help students foster (Athavale et al., 2008; Huber et al., 2005; Walker & Black, 2000). Further, where ICD approaches include a community learning component, students are able to gain valuable real-world experience, while the collaboration between the university and community stakeholders can serve to break down a 'historical lack of trust' between academia and the community (Andrées et al., 2014, p. 31). Finally, ICD allows for co-creation and collaborative approaches to curriculum design; potentially creating an environment where classroom hierarchies between students and instructors can begin to be broken down (Fenton & Gallant, 2016; Iszatt-White, Kempster, & Carroll, 2017; Lubicz-Nawrocka, 2017, 2018).

This shifting of power from the instructor as the deliverer of knowledge to a facilitator in a co-learning, experiential, and collegial environment is generally perceived as a good thing (Hibbert & Cunliffe, 2015), however there can be an attendant sense of anxiety or questioning of identity on the part of instructors (Iszatt-White et al., 2017). In her study of three different approaches to ICD, Lubicz-Nawrocka (2018) found that all of the instructors she spoke with reflected on the 'risks, vulnerabilities, and challenges' (p. 58) that are associated with a ceding of power in the classroom. One instructor described the experience as 'nerve-wracking' (Bovill, 2014, pp. 21–22).

Generating student buy-in and acceptance of creative pedagogical practice and a shift in the power dynamic can also be challenging, given that students grow to expect familiarity. Fischman and Gandin (2007) reflected on their attempts to break down the power dynamics between their students and themselves. Their students reacted with disbelief based on their own educational experiences and 'their lack of knowledge about other ways of teaching and learning and their difficulties of imagining that other school is possible' (Fischman & Gardin, 2007, p. 210). As such, students' traditional educational experiences leading up to university may influence how they may react to creative approaches. On the one hand, they may resist alternative ways of learning, however, on the other hand, students may welcome change and embrace creative approaches.

ICDs are also challenging to implement because they require a great deal of effort and planning (Athavale, Davis, & Myring, 2008). Specifically, change often requires negotiation with a gatekeeper e.g. a head of department (Powell, James & Johnson, 2013). In their article entitled *With their Permission Skeptics, Resisters, and Supporters* Powell et al. (2013) reflect on the various obstacles innovative pedagogues may face in their attempts to creatively adapt curriculum. Power dynamics exist between faculty members at various ranks, thus a junior scholar may be challenged to convince their head of department or senior colleagues to accept and/or embrace that change is needed; this may be especially challenging as senior colleagues may find comfort in predictability and resist the change suggested. The bureaucratic nature of universities may prevent innovation specifically in relation to scheduling (Powell et al., 2013) e.g. a desire to blend two classes together in one classroom. Moreover, such institutional structures and resource allocation priorities may not operate in ways that support the implementation of ICDs (Athavale, Davis, & Myring, 2008; Pharr, 2000, 2003). Such barriers may impede the sustainability of an ICD in some circumstances due to staffing changes within a department (Brunel & Hibbard, 2006, p. 15). Lastly, convincing colleagues that assessments are sound, the general quality of the content in an ICD is appropriate, and 'engaging in the age-old debate of depth versus bredth' (Powell et al., 2013, p. 116) may be necessary.

Despite their many lauded benefits, ICDs remain challenging to implement in the long term and on a large scale. Next, we turn our attention to what the literature tells us about the co-creation of curriculum, and in particular what roles community partners may play in fostering university-community collaborations.

Co-creation in and beyond the classroom

With this paper we are exploring the roles and experiences of two tourism community partners in the co-creation of an integrated curriculum initiative at the undergraduate level. This perspective has not yet been attended to in the tourism or leisure literature, a dialogue that we are happy to initiate. Emphasising how we cultivated relationships with our community partners, built trust, and nurtured these relationships by encouraging dialogue and feedback throughout our process elucidates the co-creation of our ICD. In so doing, we also model best practice in establishing meaningful community partnerships, which are mutually beneficial to students and society, a cornerstone of our departmental and faculty vision. Literature on in-campus-community collaborations often identifies community partners as the ultimate beneficiary of the research (Andrées et al., 2014). However, our efforts to involve our partners in co-creation establishes an equal partnership in our ICD. It is our intention for other scholars to build on our work reflecting on mutually supportive community partnerships in order to make the most impact to the communities in which our universities are embedded. Based on our experience with our ICD, we propose that community partners should be both actively involved in the planning *and* delivery of the ICD, and yet we should not lose sight of their needs as industry stakeholders. In this section, we present a summary of the literature dedicated to curriculum co-creation with students, followed by the roles communities and universities can have in supporting and elevating one another's missions.

Fenton and Gallant (2016) inform us that academic discourse today supports the placing of the community 'at the heart of student learning' (p. 1). Community settings are particularly well-suited for integrated and experiential learning because the world outside of the classroom is necessarily interdisciplinary (Fenton & Gallant, 2016, p. 2). Integrated learning experiences are of increasing importance in today's context of 'supercomplexity' (Lubicz-Nawrocka, 2018, p. 59), experiences that encourage students to explore the interconnected nature of the subjects that they are studying, rather than sticking to traditional disciplinary silos (Anderson, 2013; Fenton & Gallant, 2016; Huber et al., 2005; Paisley, Spencer, Wells, & Schwab, 2013). For students to gain the most out of these experiences, and begin to see themselves as drivers of their own learning, ICD approaches prioritize placing students at the heart of the design.

Traditional approaches to education have been critiqued for treating students as 'empty receptacles to be filled with their teacher's knowledge' (Smith et al., 2016, p. 28; Anderson, 2013; Bergmark & Westman, 2016; Dewey, 1986; Fenton & Gallant, 2016; Freire, 1970). Current educational theory supports the idea that students learn better when they are actively involved in their own learning, that taking initiative allows them to fully develop their potential (Bergmark & Westman, 2016; Bilous et al., 2018; Smith et al., 2016; Voogt et al., 2011). Co-creation of the curriculum allows students to take responsibility for their own academic achievements, as well as course outcomes, and opens up dialogue in the classroom about approaches to learning that can help to break down traditional classroom power structures (Lubicz-Nawrocka, 2018, pp. 47–48). When instructors cede some of the control in how the classroom operates, students experience greater satisfaction and sense of personal achievement in their academic experiences (Lubicz-Nawrocka, 2018). This in turn helps students to develop and foster the 'cognitive, interpersonal, and intrapersonal abilities' that have become increasingly essential in today's interconnected world (Lubicz-Nawrocka, 2018, p. 47). In her study of three ICD classrooms, Lubicz-Nawrocka found three factors to be central to curriculum co-creation: 'respect for different opinions, reciprocity by sharing different (although not necessarily the same) expertise and perspectives, and responsibility shared amongst students and staff' (p. 58). These factors are both 'foundational prerequisites [...] as well as outcomes' as they are fundamental to successful implementation in addition to being strengthened over time (Lubicz-Nawrocka, 2018, p. 58). Co-creation may also lead to more democratic learning environments, which, as mentioned above, may be difficult for some to see that balance of power shift, however, can also help students to prepare for more democratic and collegial working relationships in the post-school world (Bovill, 2014; Bovill, Morss, & Bulley, 2009; Bron, Bovill, & Veuglers, 2016; Lubicz-Nawrocka, 2018). Unlike in traditional learning environments, students are viewed as 'motivated *partners* in collaborative inquiry' (Bovill, 2014, p. 16, *emphasis added*). Bovill (2014) describes a liminal moment in which students realize that their instructors are listening to them and acting on their ideas. Students were also found to carry these experiences forward, as their expectations of education and their empowerment as co-creators of the learning environment grew (Bovill, 2014).

Despite a plethora of research supporting the positive educational and personal development outcomes of co-creating curricula for students, it is still comparatively rare in practice (Lubicz-Nawrocka, 2018), in part due to it being a somewhat 'nerve-wracking' experience for instructors (Bovill, 2014, p. 21, 22). Even more rare are instances in the literature of educators co-creating curricula with community partners. Up to now, the emphasis in the literature has been primarily on service learning, integrative learning, and community-based research when discussing the relationships between academia and the community. Also referred to as 'community-engaged scholarship' (Andrées et al., 2014, p. 29), the emphasis here is on ensuring mutually beneficial experiences for both students and community stakeholders (Andrées et al., 2014; Fenton & Gallant, 2016, p. 4). The idea is to provide students with meaningful learning experiences that put theory into practice, while at the same time giving back and strengthening the community (Andrées et al., 2014; Brzuzy & Segal, 1996; Ward & Wolf-Wendel, 2000). Students develop work-relevant skills and learn about real issues in the community, while community stakeholders get to leverage the resources of the university, such as research and volunteer hours, that they might not otherwise have access to on limited budgets (Andrées et al., 2014; Brzuzy & Segal, 1996). One clear advantage that has been identified is that students conduct research that might not otherwise be done (Brzuzy & Segal, 1996, p. 67). This leads to more information being collected by the university, and the community organisation learning more about its client population and environment, which in turn can lead to social change (Brzuzy & Segal, 1996; Chopyak & Levesque, 2002).

Much of the emphasis on service or community-based learning has been focused on the *doing for*, a perspective more to do with charity than with social change (Ward & Wolf-Wendel, 2000, p. 769). Universities in North America have been criticised for distancing themselves from the communities in which they are embedded (Goss, Gastwirth, & Parkash, 2010), and where they have made efforts to be more involved in the community, community partners are more frequently viewed as beneficiaries or objects of study rather than as partners in research and learning (Goss, Gastwirth, & Parkash, 2010; Ward & Wolf-Wendel, 2000). For Chopyak and Levesque (2002) the benefits of creating true community-university partnerships are multiple. Communities are facing more problems than they can effectively cope with on their own and are able to leverage university resources, university administrators are able to demonstrate how they are adding value to the community, and students have an opportunity to develop 'real-world' skills that today's employers are seeking (p. 204).

Despite a lack of attention paid to the role of community partners in the co-creation of an ICD at the post-secondary level, in this section we have outlined the ways in which co-creation has been found to have had positive impacts for student participants, as well as to the desirability of true university-community partnerships.

Methodology

Our previous scholarship has qualitatively reflected on our ICD development from an undergraduate student experience and graduate student experience (Boluk et al., 2019). As stated at the outset, the aim of this paper is to investigate the process of co-designing and co-constructing an ICD alongside our community partners. Data for this project is grounded in in-depth semi-structured interviews collected from April to July 2018 following our ICD implementation. Members of the research team carried out one-on-one interviews with the two instructors involved in the ICD delivery, two of our five community partners (who were available and responded to our request), and six teaching assistants. Each interview lasted approximately 60–90 minutes and was audio-recorded and transcribed. Questions guiding these interviews included: What was your involvement in the ICD project? What were your reactions to the integrated process and the assignments? What did the outcomes of the students' work mean to you? What was your general experience with our ICD project? How was your involvement in our ICD similar or different to your other experiences teaching/as a TA/ or how did it compare to other projects you have been involved in with our third level students? Was there anything particularly meaningful from your involvement in ICD? At this latter stage of analysis, we decided to use Creative Analytic Practice as a way to communicate findings in an engaging and accessible way for both academics and industry partners alike.

Creative analytic practice

Creative Analytic Practice (CAP) treats the process of writing itself as an important tool in the analysis of data, considering information from new perspectives. Taking a CAP approach means that the researcher understands the '*writing process* and *writing product*' to be inextricably intertwined (Richardson, 2000, p. 930, emphasis in original). CAP clearly foregrounds the perspective of the writer, however, the reader is left to draw his or her own conclusions as to the meaning of the events that are portrayed (Berbary, 2011; Parry & Johnson, 2007; Richardson, 2000). Founded in the postmodernist and poststructuralist movements, CAP demands that researchers move beyond the staid, impersonal, 'god-trick' of traditional writing in the social sciences, to experiment with writing that is lyrical, personal, embodied, and emotive – narratives that van Maanen would describe as 'impressionistic tales' (1988). A number of authors, including Vannini (2015), Lorimer (2005), Richardson (2000), Höckert (2015, 2018), and Thrift (2000, 2008) look to non-representational theory in order to go beyond representation in order to more effectively encapsulate people's variegated lived experiences, without succumbing to the temptation to translate them through their own frames of reference and analysis. To Richardson (2000) CAP is 'writing as a method of inquiry, a way of finding out about yourself and your subject' (p. 923). CAP texts can be messy, poetic, evocative, mysterious, whatever form the author has found that allows them to engage deeply with the material and reflect upon its meaning and import (Berbary, 2008; Done et al., 2011; Parry & Johnson, 2007; Richardson, 2000).

In choosing to apply CAP, researchers are acknowledging and laying bare the values, ideologies, biases, and standpoints that play a crucial role in informing how we are able to know what we know (Richardson, 2000). In her work on the subject, Richardson (2000) bemoans the dull, dreary, and uninspired writing up of so much qualitative inquiry. CAP allows writers to play with their writing style, and in so doing uncovering new perspectives while also creating texts that are appealing to readers, including those outside of academia (Berbary, 2011; Denzin & Lincoln, 2003; Richardson, 2000). Rather than attempting to condense the findings of a study to have it accord and 'fit in' with an existing theoretical framework, CAP 'opens up the research and invites questioning and ongoing interpretation (Berbary, 2011; Boluk et al., 2019; Richardson, 2000). CAP 'encourages involvement, inspires curiosity, creates inclusivity, and constructs depictions that remain in the thoughts of readers in ways that traditional representations sometimes do not' (Berbary, 2011, p. 194).

The fact that CAP emphasises the production of creative and aesthetically appealing texts, minus the academic jargon that is often an immediate barrier to lay audiences, means that the research that is being shared can potentially have a much broader impact in society. Readers want to be able to relate, to see themselves in the narratives that are being told (Berbary, 2011; Grybovych & Deiser, 2010). This is one of the means by which researchers are able to 'change the world by the way they make it visible' (Parry & Johnson, 2007, p. 124). Within the growing body of work that characterises itself as 'art as research, research as art' (Berbary, 2011, p. 194), CAP allows researchers to produce political, evocative, collaborative, creative, decolonising, and honest reflections on and of our social world.

CAP has received scant attention in the field of critical tourism studies, however those studies that have explored its potential are noteworthy. In the classic text *The body and tourism* by Veijola and Jokinen (1994) our narrator and her travelling companion share their disappointment with traditional cerebral approaches to understanding tourism that completely ignore the role of the body through the use of their imagined encounters with some of tourism's most notable scholars at a beach resort. In her work with rural tourism providers in Nicaragua, Höckert (2015, 2018) uses creative writing techniques to trouble dynamics of power and issues of representation that arose over the course of her ethnographic work. Wright (2018) employs vignettes as a way to provide insights into the experiences of ageing sport tourists. Lorimer (2013) travels the Scottish countryside in search of the increasingly scarce scarecrow and tells of his adventures in an essay which 'operates as a cipher for stories old and new, of agricultural society, country life, landscape politics, and rural values' (p. 177). Jensen et al. (2015) look to non-representational theory to present a series of 'audio-visual impressionistic tales' (p. 62, and informed by van Maanen) as well as netnography to explore experiences of interrailing in Europe.

Given the opportunities CAP presents it is surprising that more tourism scholars have not taken up this representational strategy. In our view, CAP is an opportunity to engage in social justice as the stories we share representing the experiences of a teaching team, equally paying tribute to a junior scholar, a seasoned scholar, graduate teaching assistants, and a two community partners involved in co-creating our ICD initiative. We co-constructed our vignettes following a close reading of our transcripts, reviewing notes taken in meetings, and reviewing emails exchanged among our research team. Our choices in the findings section attempt to creatively narrate and blend our data to reflect the larger data set, but with findings focused specifically on our relationships to industry partners.

Findings

Our analysis of the data revealed several opportunities in consciously studying our pedagogical practice. Specifically, our themes reflect the impact of building a supportive network in order to meaningfully engage in pedagogical scholarship and practice. Our analysis also reveals the prospects of involving community partners, learning from them, and co-constructing activities based on the reflections of all those involved.

Realising possibilities and making connections

The initial idea of our ICD was an outcome of Karla and Corey meeting and determining their mutual interest in pedagogical scholarship, carrying out an initial pilot study supported by a small seed grant and then a larger, more comprehensive grant to consciously and deliberately study our pedagogical practice.

> My new sneakers squeak as I jot up the steps clutching my cold brew coffee. As I hit the top step, bright light floods the lobby of the newly constructed Nano-Tech building where I was meeting Karla. Surveying the open space, an array of almost equal amounts of white, brown, black, and Asian students scattered about on sofas or chairs, most with their heads buried in mac laptops, I don't spot her. This landscape, this culture, this school is still so new to me. I pause, thinking to myself, 'I thought I was late. Wait, am I lost again?' It had happened so many times before. A year on campus and I still had a hard time finding my way around, both physically and culturally. Although I had almost a 20-year career as a faculty member across three different institutions in the U.S., the University of Waterloo was quite different. As the most innovative

university in Canada, STEM was highly valued over other disciplines and the high achieving students were used to and seemed to demonstrate a preference for standard lectures. In addition, as a research-intensive University, teaching was seen as 'something you just need to do, so you can focus on your research.' I had heard that before at my last institution. However, as someone who values and thinks a lot about pedagogy, I had always found a team to work with on curriculum development, scholarship of teaching and learning projects, or the mentoring of junior faculty. But being a thoughtful pedagogue wasn't going to be enough, I didn't really know this community and despite my desire to connect our students to 'real-life' experiences, I was ill prepared. Just as I was starting to question my move here, I spotted Karla across the room. In jeans, a blue jacket and scarf, with soft brown hair that hit her shoulders and glasses that slightly pulled her hair from her face to reveal a warm smile. I waved and walked her way. I felt a bit remiss that I hadn't really gotten to know Karla yet. I was meeting her today because during my performance review with my department head, I had shared that I had experienced the lowest teaching evaluations of my career and I suspected much of that was due to trying to implement experiential education in a community that was not really 'mine' yet. 'No need to worry as a full professor, your research productivity is right on target' he said. 'I know, but I am also interested in good instruction,' I said. I began to leave the office a little disillusioned and he could read it on my face. As I crossed the threshold of the door, he yelled after me. 'You know Karla has a strong interest in teaching and working with community. I think she even writes about it.' Maybe Karla was who I was looking for? Someone who I could share my passionate for pedagogy and help me learn more about the local community and Canadian context, but also someone who might benefit from my desire to be a mentor. Little did I know, in many ways, she would be mentoring me.

Building on strengths and developing partnerships

From: Karla Boluk
Sent: 24 September 2017 19:42
To: Corey Johnson
Subject: Meeting to discuss an ICD?

Hiya Corey,

Would you believe I just got a call from the CEO of our regional tourism marketing organization (RTO4)? He asked to set up a meeting to discuss some possible synergies. Specifically, he is interested in exploring the digital sophistication of tourism businesses in four municipalities. I mentioned my interest in working more closely with industry stakeholders so we may design meaningful assessments for our students that are rooted in our community. I reiterated the caliber of our students ... he was well aware as an alumnus, who derived benefits from our co-op programme. Our co-op programme is fantastic but it must be such an adjustment for our students to return to the classroom in between their placements. I wonder how we may build on our students' experiences, and encourage them to carry forward what they learned in previous co-op terms? Imagine if they were able to teach their peers and profs what they learned, and apply practice to theory and vice versa! It would be great if we could work more closely with the industry so there is some continuity between students' co-op placements and their structured classes. I'd love to see our students provide some tangible outcomes to our partners, after carrying out some research, this may make a nice capstone project and also create a synergy between their theoretical and practical learnings.

I suppose this ICD notion is really starting to percolate. Would you have time for a coffee late next week after my meeting with RTO4?

Cheers!

K

Karla Boluk, Ph.D.
Assistant Professor,
Department of Recreation and Leisure Studies
Faculty of Applied Health Sciences
200 University Avenue, University of Waterloo

I adjusted my posture in the brown, worn-in, comfy leather chair positioned in the corner of the University Starbucks. I was anticipating filling Corey in on all of the details of my conversation with the CEO of RTO4. I was delighted they reached out, and excited they brought forward a timeless problem. I was hoping our undergraduate students would share my enthusiasm, and hopefully this was a project that would help build some partnerships.

Corey peaked around the door and pulled up the only remaining chair in the faculty atrium bordering the café. While getting comfortable in his chair he asked, 'How are things?' Keen to learn about my conversation with RTO4 he prompted 'I'm looking forward to hearing about your dialogue'. Attempting to cut to the chase I highlighted that I had met the whole team at the Stratford office and they were all very lovely and keen to explore ways to collaborate. I was impressed they found me given that I imagine universities are not easy to navigate from the outside. We discussed a research project RTO4 was interested in getting off the ground on the 'Digital Sophistication of Tourism businesses' in four municipalities. This conversation segued into some problems RTO4 was aware of facing tourism businesses in the region. We spent the remainder of the meeting talking about such challenges, me, with my pedagogy cap on trying to determine how I could turn these 'problems' on their heads via student engagement.

'Were there any problems which really stood out?' Corey asked attempting to narrow in on one overarching question which may give our ICD focus. 'Yes! There was one concern that stood out to me, in regard to being more in tune with the millennial user experience'. One of the concerns RTO4 articulated was that many tourism businesses would benefit from learning more about what may enhance the millennial experience. Corey perked up and said immediately 'this is fantastic, we can help identify the millennial UX given our direct access to students, this is exactly what we need to frame our ICD, great work!' I nodded. (Karla)

From building partnerships to sustaining interest

From: Waterloo Regional Museum
Sent: 16 April 2018 10:22
To: Karla Boluk
Subject: Collaborating with students

Dear Karla,

Thanks so much for partnering with the Waterloo Regional Museum. We approached this project with the notion that we had nothing to loose, you made it very easy for us, our involvement wasn't burdensome at all. We were attracted to the project because we know we have not been on the radar of the student market so we were looking forward to getting some valuable feedback from this group regarding how we may be able to reach them. Another attractive element of this assignment was that the students coming in, didn't self identify as "being part of the project", so we weren't judged based on our effort to put our best face forward. The feedback was realistic, timely and authentic. To be honest, I wasn't expecting the scope to be as broad as it was, encompassing the full user experience, ranging from the front desk and gift shop through to the exhibits. I anticipated it would just be about the museum galleries. Having said that, everybody knows people will judge you by the cleanliness of your washrooms, so its nice to see the students' eye for detail. The students' work surpassed my expectations.

The direct student feedback is what we needed and this is exactly what we received. Having said that there was some additional content that went beyond what we needed to know such as the background of what has been documented in the museum literature world. But I knew what I was looking for and so I was able to read around the content to pick out what I needed. The final product was more of an essay, all I really needed to know was "here are the questions we asked, here are the answers, and here are the top three suggestions". That would have been a little more concise than reading through a whole essay. At the end of the day, I want to be able to share these findings with my coworkers or boss, and its hard to tell her "can you read this 27 page document"? and you know filter through to find the evidence to support making change X? If I gave her the key bullet points that the students noted that's a little more sharable, and digestible and takes the task away from me to do it.

One aspect of the project I have been deliberating on is the role of the Expert Users, your first year students. I am not entirely sure what the thirty students go tout of the project because I really didn't interact with them at all but, I would like to! It was the fourth year students who directly reached out to me. Do the first year students see a write up of the outcomes in the end to tie up their time spent in the project? I noticed in the final presentations there wasn't too much engagement with the first years. I think its often the case when it comes to presentations, you know you prepare, get up and present and you wish there were more questions and more involvement from the audience sitting there. Luckily there was one gentleman on the panel who was asking a good bunch of questions rolling from the first years. I'd almost like to get the 30 people who visited together in a room to have that targeted audience at my disposal for a little while and pick their brains, you never know when you will find your next best idea. I would then give these 30 students free passes that were marked in some way, just to see if they would come to our museum again. If they didn't use them I think we could take that as an indication that we were still missing the mark on something.

p.s. As indicated previously we probably employ about 30 co-op students every summer so we have had first hand experience with your students! It would be neat to see more student from Recreation and Leisure Studies apply for a co-op at our museum, we have been asking ourselves "where are they?" maybe working together in this project will draw some of your students our way, since some of them will be going into this profession.

Keri

Guest Relations Manager
Waterloo Regional Museum

From: CTRL V
Sent: 3 April 2018 9:50
To: Karla Boluk
Subject: Student Presentations

Dear Karla,
I just wanted to drop you a line to say that Ryan and I were really happy with the ideas the students presented last week. I also wanted to follow up because you had asked that we reflect on our experiences working together over the last few months and consider how we may build on our relationship heading into the future.

Ryan and I thought your curriculum development was a fantastic way to get some on the ground feedback that we may be able to take action on. We were impressed from the outset with the professionalism of the two students responsible for our attraction. First, in their email requesting we engage as an attraction and then in our meeting when they showed us their User Experience Survey they planned to pass onto the expert users. I was pretty impressed with the type of questions they developed and I provided feedback on the initial draft of their survey to tailor it appropriately to our clients. I must say though I am not entirely sure if they took up my suggestions because I didn't see a final draft of the survey before they sent it out. We have some ideas regarding how we may build on the current framework. We really feel like we have more to offer your students! While it may not have been the intention, we were left out of the loop until the final presentation a couple of months after we were approached to be involved and so we had some questions. I have a few ideas in mind. I mentioned previously, I'm not sure if the students implemented my suggestions to their survey so I would have liked to see their final draft. Then, we didn't get any of the results back. We got an overview in their final presentation but, I would have loved to actually see the raw data. And I would have liked to work with the students more closely on their presentation. We felt like we could have taught the students more about our brand and what we are trying to do here. Generally, I feel like we could have provided a lot more insight. The final presentation was a great general overview but what I have been thinking about is the space, the presentations were held in a very large lecture theatre, so some of the first year students were checked out! I get it, I remember being a first year … no offense! Perhaps it would be better if we had smaller groups of students, what if the final presentation was a boardroom style type of presentation? What if we hosted it here at our venue? We are just across the street after all. A smaller group may help encourage dialogue and feedback.

Carly
CTRL V, Waterloo

Key learnings for students from industry's perspective: desired involvement & meaningful student products

Following class Corey was making his way back to his office when he noticed Karla's office door slightly ajar. He popped his head in the door and said, 'how have the interviews with our community partners been going?' I took a peak at the transcripts in our Dropbox and I think we have some great data here. 'You have some time?' he asked. She nodded her head vigorously, as he slid in the door occupying the only chair at the other end of her desk.
'Corey, you wouldn't believe the feedback we have received!' Overall our partners are genuinely happy with the products our students have produced and they have some really helpful solutions to help us improve our work'. First of all, I am inspired by the fact that our partners want to continue working together! One of our partners stated that 'It was neat to see a project whereby the students are mentoring themselves, and we acted as a resource'. I thought this signalled that we are alleviating the burdensome nature of what some industry-classroom relationships may look like and both community partners retold stories of partnerships that were either taxing or did not deliver so this is important for us to remember moving forward.
Surprisingly, they came forward with suggestions regarding how to enhance their engagement- as in they are actually willing to give more of their time! Without my specifically asking of this I was delightfully surprised by this! Corey stood taller in his seat 'this could generate further opportunities from a developmental perspective for our first-year students, as they practice articulating their critiques in a professional and constructive way to a community partner.' Indeed. Karla reached for her peppermint tea for a quick sip before starting to speak again. We also had a useful discussion about what a more valuable end product may look like, their suggestion was to swap out the essay for an industry friendly package. They seemed interested in a concise summary of the outcomes of the data analysis, a constructive trip review from the 1st year students, presentation slides, and possibly a report replacing the essay assessment. I don't know why this didn't occur to me beforehand! 'This recognizes

the importance of consciously reflecting on our pedagogical practices! Sounds like you have gleaned some helpful insights which is great, did you ask our community partners if they had implemented any of the students' ideas?'

Yes I did! I was delighted to learn that some of the ideas presented by our students have been implemented. Interestingly, both community partners mentioned that there were some new nuggets that gave them a fresh way of looking at things and then some of the ideas presented by our students seemed to be confirmatory but framed in a way that gave the ideas more authenticity. Seemingly, our students had some helpful advice on the use of social media and the general presentation of their businesses on-line. Waterloo Regional Museum said for example, 'there were specific suggestions regarding how they may improve our social media recognizing a shortfall with their Snapchat filter. Similarly, CTRL VR stated she had benefited from the suggestion to make some changes to the [local attraction 2] social handles, because their head office had been acting as the Waterloo site social handles. So, the students' suggestions seemed to lead to them taking action on a decentralized marketing methodology. These changes fed into a larger website overhaul including another suggestion regarding the ability for customers to leave review content on their website so [local attraction 2] added a button that says, 'leave us a review on google', because students indicated this didn't exist. Seemingly, receiving feedback from 30 students gave our partners the evidence they needed to implement the changes made. Both partners did say that some of the recommendations made won't be onboarded because of the way their companies are set up, they just weren't applicable, which underlines the opportunities in working more closely with our partners.

Key learnings for students from industry's perspective: mutually beneficial experiences

Ultimately, the interviews we carried out illustrated mutually beneficial experiences for community partners and students alike. Waterloo Regional Museum noted the final presentations 'really gave [student 1] and [student 2] the time to shine and present the findings they worked so hard on'. Members of our teaching team said for example:

> I was impressed with what the undergraduates achieved over the term and the contributions they made to our community partners. It seemed like what businesses were missing is the information they needed to adopt the right approach in some respects. So these inputs they received from the presentations and communication with students is what they needed [...] its was a win-win. (TA 3)

and 'it seemed like they [community partners] were surprised in a positive way, hearing where they were lacking so they can maybe adjust or you know fix it' (TA 4), 'I think they were shocked about how good the feedback was' (TA 5). Interestingly, TA 3 emphasised the benefit of student consultants particularly for smaller businesses who may not have the budget, support, or skills to investigate breakages in their user experience. As such, students' ideas were welcomed and perhaps came across as even more 'valuable.' Recognising the impact for small businesses was important and determined to be a pathway forward mutually recognising the contributions of our programme and university within the community and ways to give back. He noted the level of satisfaction among our partners, saying, 'people were happy, they were very encouraging about their experience, and they are interested to have more [engagement] in the future.'

A similar insight was offered by another member of our teaching team who reflected on the 'tremendous value in this assignment':

> The 4th year students had the opportunity to be in a leadership position; they led discussions, presented to destination representatives and a large group of students, and had the opportunity to participate in research. This will look great on a resume, and I think has the potential to encourage undergrads to continue school at a graduate level. I think the 2nd year students had the opportunity to be introduced to the research process, and be a part of learning in a real life context. Their recommendations were heard by the 4th year students and our community partners. They got the chance to see the 4th year students in this leadership position, and potentially it could encourage these students to aspire to complete this project when they are at the same point in their degree. (TA 6)

Presentations were followed by a networking lunch for the undergraduate presenters, the industry panel, and graduate students. One member of our teaching team remarked on this saying:

> Aside from the work required to organize and manage students, I noticed that the instructors had built relationships with the industry and with a number of business partners. I liked this aspect because I came from the Tourism Industry before I started grad school. I was a tour guide in the United Arab Emirates. I appreciated the invitations to the networking lunches following the student pitch presentations at the end of the term because I got to connect with those who work in tourism here. It's challenging to build networks with the community specifically, because students are often deemed transient. (TA 3)

Discussion/conclusion

Our data draws attention to the various benefits of consciously reflecting on our pedagogical practice. As stated by Phi and Dredge (2019) in the introduction to this special issue, co-creation is a movement towards reflexive and intentional processes, supporting creativity and the deconstruction of traditional binaries. This initiative supports collaboration across faculty hierarchical designations, as well as with community partners, with the ultimate intention of creating an innovative and engaging learning environment for undergraduate students. As this case demonstrates, collaboration requires planning and compromise and, in its best instance, may lead to the building of trust between junior and senior faculty members, as well as between academia and the community. Looking beyond our individual case, this experience suggests to us that co-creation supports the establishment of 'shared responsibility, public-private partnerships, and capacity-building' (Phi & Dredge, 2019). As such, our ICD provides an example of an innovative, transactive process that brings value beyond the tourism classroom, while at once also teaching us how to improve outcomes for both students and community partners. Consciously reflecting on the process of continuously designing our ICD allowed us to learn how we could better serve the needs of our community partners.

Greater involvement across all partners in the ICD, as well as increased opportunities for communication can help the students to develop feelings of responsibility and self-sufficiency in problem solving, in addition to bringing university and community partners together to solve complex societal problems (Andrées et al., 2014; Athavale, Davis, & Myring, 2008, p. 295; Bilous et al., 2018; Bovill, 2014; Fenton & Gallant, 2016; Lubicz-Nawrocka, 2017, 2018). Our data revealed that both community partners felt they could enhance student benefits through increased engagement with them and involvement in the course planning processes. We also have additional work to do in order to ensure that the final results of the project are adapted to the needs of our community partners. In accordance with common evaluation practices at our university, the students were asked to prepare a final essay summarising their work over the course of the ICD. However, this format was not suitable for our community partners, who would have preferred a succinct final report, which would have been sharable with their colleagues to promote the recommended changes. As such, there was a desire for a more authentic assignment that was industry connected and would thus streamline the final content. Now that we have this information, in the future we will be cognizant of our community partners' output needs and encourage greater involvement on the part of community partners in the planning of this part of the students' assessments, and as well as in its evaluation in terms of its usefulness to an industry audience. The literature suggests that in campus-community collaborations, the community is too often treated as a beneficiary of the research, rather than as equal partners in its design (Andrées et al., 2014; Ward & Wolf-Wendel, 2000). Based on our experience with our ICD, we propose that community partners be both actively involved in the planning *and* delivery of the ICD, and yet we should not lose sight of their needs as industry stakeholders.

Our community partners also reflected on the ways in which our tourism students were particularly well suited to meet their objectives in aligning themselves with this ICD. Where previously the local museum had worked with students from the Faculty of Arts, these students were found to be more interested in the content of the exhibits, whereas the tourism students were focused on the entire UX and thus applied a broader lens to assess all aspects that comprise the museum. This insight may have an unintended downstream consequence, as, through increased contact, community stakeholders develop a more intimate understanding of what graduates of our programme are capable of.

The presentation of our data reveals the central role of everyday practices in developing a project of this nature. For example, our initial vignette reflects on a performance evaluation debrief meeting between Corey and our department chair. Corey's expressed concerns regarding his low teaching evaluations led to a whimsical comment made by our chair who suggested Karla's mutual interest in pedagogy and pedagogical scholarship. This led to an initial meeting where mutual interests were established, Corey planted the seed about ICD, and a collegial partnership was born. Collegiality was particularly salient in the development of our ICD project given that Karla often sought Corey's seasoned advice and Karla's TAs required clarification at times to provide clarity to the students. From there Karla was able to see a place in her curriculum for a co-ordinated response to the community problem presented by RTO4 regarding improving their understanding of the millennial user experience. The research team encouraged dialogue throughout the ICD as such emails became an important communication vehicle and way to receive feedback. Inviting our community partners to critique our ICD was important because as we noted in our vignettes our partners came forward with suggestions regarding enhancing student engagement, and providing more meaningful products. This feedback received from our community partners led to the true co-creation of our ICD. Furthermore, the investment of our community partners was apparent when they volunteered to be co-educators which we believe will enhance the experiences of all those involved. Our vignettes also display an emotional journey on behalf of those involved especially when developing something new from the ground up.

Our experience in conducting an ICD with community partners was not without its challenges. Partnering and developing content that is both of interest and educational value to the students *and* of ongoing benefit to community partners requires a great deal of effort, energy, communication, and planning. University term schedules are frequently fixed and not easily changed once rooms are allocated and syllabi are finalised, a rigidity which may not work for smaller or more nimble local partners. It would also be desirable, in order to enhance the overall experience, to include community partners as co-instructors, either in the classroom or in a community setting (Fenton & Gallant, 2016). This, however, is only possible once a rapport has been established through experiences of working with the students and university instructors for our community partners to probe and benefit more fully from the exercise.

This paper follows a previous publication (Boluk et al., 2019) analysing the process of designing and delivering an Integrated Curriculum Design project through the application of a Creative Analytic Practice. CAP is a novel approach to creatively presenting data with the hope of broadening the reach of the study beyond the traditional academic audience. While not widely used in our field, CAP has the potential to enrich future studies exploring the opportunities of higher education. Future research will look more closely at the experiences of student participants in an ICD, including those of the student graduate teaching assistants who collaborated as part of the teaching team. This work will specifically reflect on the significance of an intergenerational teaching team involved in developing our ICD. In conjunction with reflections on intergenerational teaching teams, we will reflect on the experiences of co-creation along with our community partners.

Disclosure statement

No potential conflict of interest was reported by the authors.

Funding

This research was supported by the University of Waterloo's Learning Innovation and Teaching Enhancement Grant.

ORCID

Corey Johnson http://orcid.org/0000-0002-6918-3787

References

Ackoff, R. L., & Greenberg, D. (2008). *Turning learning right Side Up*. Upper Saddle River, NJ: Prentice Hall.

Anderson, D. (2013). Overarching goals, values, and assumptions of integrated curriculum design. *Schole: A Journal of Leisure Studies and Recreation Education*, *28*(1), 1–10.

Andrées, P., Chapman, D., Hawkins, L., Kneen, C., Martin, W., Muehlberger, C., ... Stroink, M. (2014). Building effective relationships for community-engaged scholarship in Canadian food studies. *Canadian Food Studies*, *1*(1), 27–53.

Athavale, M., Davis, R., & Myring, M. (2008). The integrated business curriculum: An examination of perceptions and practices. *Journal of Education for Business*, *83*(5), 295–301.

Baldwin, E., Mainieri, T., & Brookover, R. (2013). The EDGE of learning. *SCHOLE: A Journal of Leisure Studies and Recreation Education*, *28*(1), 64–77.

Berbary, L. (2008). *Subject to sorority: Women's negotiations of competing discourses of femininity*. Doctoral dissertation, University of Georgia. Retrieved April 2, 2014 from https://getd.libs.uga.edu/pdfs/berbary_lisbeth_a_200808_phd.pdf.

Berbary, L. (2011). Poststructural Writerly representation: Screenplay as creative Analytic practice. *Qualitative Inquiry*, *17*(2), 186–196.

Bergmark, U., & Westman, S. (2016). Co-creating curriculum in higher education: Promoting democratic values and a multidimensional view on learning. *International Journal for Academic Development*, *21*(1), 28–40.

Bilous, R., Hammersley, L., Lloyd, K., Rawlings-Sanaei, F., Downey, G., Amigo, M., … Baker, M. (2018). 'All of us together in a blurred space': Principles for co-creating curriculum with international partners. *International Journal for Academic Development, 23*(3), 165–178.

Boluk, K. A., Muldoon, M. L., & Johnson, C. W. (2019). Bringing a politics of hope to the tourism classroom: Exploring an integrated curriculum design through a creative and reflexive methodology. *Journal of Teaching in Travel & Tourism, 19*(1), 63–78.

Bovill, C. (2014). An investigation of co-created curricula within higher education in the UK, Ireland, and the USA. *Innovations in Education and Teaching International, 51*(1), 15–25.

Bovill, C., Morss, K., & Bulley, C. (2009). Should students participate in curriculum design? Discussion arising from a first-year curriculum design project and a literature review. *Pedagogic Research in Maximizing Education, 3*, 7–26.

Bron, J., Bovill, C., & Veuglers, W. (2016). Students experiencing and developing democratic citizenship through curriculum negotiation: The relevance of Garth Boomer's approach. *Curriculum Perspectives, 36*(1), 15–27.

Brunel, F., & Hibbard, J. (2006). Using Innovations in student Teaming to leverage Crossfunctional and Marketing learning: Evidence from a fully integrated undergraduate Core. *Marketing Education Review, 16*(3), 15–23.

Brzuzy, S., & Segal, E. A. (1996). Community-based research strategies for social work education. *Journal of Community Practice, 3*(1), 59–69.

Chopyak, J., & Levesque, P. (2002). Community-based research and changes in the research landscape. *Bulletin of Science, Technology & Society, 22*(3), 203–209.

Denzin, N., & Lincoln, Y. (2003). *Strategies of qualitative inquiry* (2nd ed.). Thousand Oaks, CA: Sage.

Dewey, J. (1986). Experience and education. *The Educational Forum, 50*(3), 241–252.

Done, E., Knowler, H., Murphy, M., Rea, T., & Gale, K. (2011). (Re)writing CPD: Creative analytical practices and the 'continuing professional development' of teachers. *Reflective Practice, 12*(3), 389–399.

Fenton, L., & Gallant, K. (2016). Integrated experiential education: Definitions and a conceptual model. *Canadian Journal for the Scholarship of Teaching and Learning, 7*(2), 1–17.

Fischman, G. E., & Gandin, L. A. (2007). Escola Cidada and critical Discourses of educational hope. In P. McLaren & J. L. Kincheloe (Eds.), *Critical pedagogy where Are We Now?* (pp. 209–221). New York: Peter Lang.

Freire, P. (1970). *The pedagogy of the oppressed*. New York: Herder and Herder.

Goss, K. A., Gastwirth, D. A., & Parkash, S. G. (2010). Research service-learning: Making the academy relevant again. *Journal of Political Science Education, 6*, 117–141.

Grybovych, O., & Deiser, R. (2010). Happiness and leisure: An ethnodrama, Act I. *Leisure/Loisir, 34*(1), 27–50.

Hibbert, P., & Cunliffe, A. (2015). Responsible management: Engaging moral reflexive practice through threshold concepts. *Journal of Business Ethics, 127*(1), 177–188.

Höckert, E. (2015). *Ethics of hospitality: Participatory tourism encounters in the Northern Highlands of Nicaragua. Acta Universitatis Lapponiensis 312*. Rovaniemi: Lapland University Press.

Höckert, E. (2018). *Negotiating hospitality: Ethics of tourism development in the Nicaraguan Highlands*. New York: Routledge.

Huber, M., Hutchings, P., & Gale, R. (2005). Integrative learning for liberal education. *Peer Review, 7*(4), 4–7.

Iszatt-White, M., Kempster, S., & Carroll, B. (2017). An educator's perspective on reflexive pedagogy: Identity undoing and issues of power. *Management Learning, 48*(5), 582–596.

Jensen, M. T., Scarles, C., & Cohen, S. A. (2015). A multisensory phenomenology of interrail mobilities. *Annals of Tourism Research, 53*, 61–76.

Klein, J. T. (2005). Integrative learning and interdisciplinary studies. *Peer Review, 7*(4), 8–10.

Lorimer, H. (2005). Cultural geography: The busyness of being `more-than-representational'. *Progress in Human Geography, 29*(1), 83–94.

Lorimer, H. (2013). SCARING CROWS*. *Geographical Review, 103* (2), 177–189.

Lubicz-Nawrocka, T. (2017). Co-creation of the curriculum: Challenging the status quo to embed partnership. *The Journal of Educational Innovation, Partnership and Change, 3*(2), 1–14.

Lubicz-Nawrocka, T. (2018). Students as partners in learning and teaching: The benefits of co-creation of the curriculum. *International Journal for Students as Partners, 2*(1), 47–63.

Malik, A., & Malik, R. (2011). Twelve tips for developing an integrated curriculum. *Medical Teacher, 33*(2), 99–104.

Miller, R. (2005). Integrative learning and assessment. *Peer Review, 7*(4), 11–15.

Paisley, K., Spencer, C., Wells, M., & Schwab, K. (2013). The university of Utah's integrated Core. *Schole: A Journal of Leisure Studies and Recreation Education, 28*(1), 54–63.

Parry, D., & Johnson, C. (2007). Contextualizing leisure research to encompass complexity in lived leisure experience: The need for creative analytic practice. *Leisure Sciences, 29*, 119–130.

Pharr, S. W. (2000). Foundational considerations for establishing an integrated business common core curriculum. *Journal of Education for Business, 76*(1), 20–23.

Pharr, S. W. (2003). Integration of the core business curriculum: Levels of involvement and support provided. *Marketing Education Review, 13*(1), 21–31.

Phi, G., & Dredge, D. (2019). Editorial: Critical issues in tourism co-creation. *Tourism Recreation Research, 44*(3), 281–283.

Powell, G. M., James, J., & Johnson, C. W. (2013). With their Permission. *Schole: A Journal of Leisure Studies and Recreation Education, 28*(1), 113–121.

Richardson, L. (2000). Writing: A method of inquiry. In N. Denzin & Y. Lincoln (Eds.), *The handbook of qualitative research* (2nd ed., pp. 923–948). Thousand Oaks, CA: Sage.

Shoemaker, B. J. E. (1989). Integrative education: A curriculum for the twenty-first century. *OSSC Bulletin, 33*(2), 1–57.

Smith, K., Kulinna, P., Vissicaro, P., & Fredrickson, L. (2016). Anthropology, dance, and education: Integrated curriculum in social studies. *The Social Studies, 107*(1), 28–37.

Thrift, N. (2000). Non-representational theory. In R. J. Johnson, D. Gregory, G. Pratt, & M. Watts (Eds.), *The dictionary of human geography* (pp. 503–505). Oxford: Blackwell.

Thrift, N. (2008). *Non-representational theory: Space, politics, affect*. New York: Routledge.

Van Maanen, J. (1988). *Tales of the field: On writing ethnography*. Chicago: Chicago University Press.

Vannini, P. (2015). Non-representational ethnography: New ways of animating lifeworlds. *Cultural Geographies, 22*(2), 317–327.

VanTassel Baska, J., & Wood, S. (2010). The integrated curriculum model (ICM). *Learning and Individual Differences, 20*(4), 345–357.

Veijola, S., & Jokinen, E. (1994). The body in tourism. *Theory, Culture, and Society, 11*(3), 125–151.

Voogt, J., Westbroek, H., Handelzalts, A., Walraven, A., McKenney, S., Pieters, J., & de Vries, B. (2011). Teacher learning in collaborative curriculum design. *Teaching and Teacher Education, 27*(8), 1235–1244.

Walker, K. B., & Black, E. L. (2000). Reengineering the undergraduate business core curriculum: Aligning business schools with business for improved performance. *Business Process Management Journal, 6*(3), 194–213.

Ward, K., & Wolf-Wendel, L. (2000). Community-Centered service learning. *American Behavioral Scientist, 43*(5), 767–780.

Wright, R. K. (2018). 'Doing it for Dot': Exploring active ageing sport tourism experiences through the medium of creative analytical practice. *Journal of Sport & Tourism, 22*(2), 93–108.

OPEN ACCESS

Student living labs as innovation arenas for sustainable tourism

Eva Maria Jernsand

ABSTRACT

This paper aims to explore sustainable tourism in relation to the concept of student living labs, defined as spaces for open innovation, co-creation and experimentation in real-life settings with students. Although these features are vital for the tourism industry, living labs are seldom discussed nor used in tourism research, education and practice. To illustrate living labs' organisation, facilitation and impediments, the author uses an ethnographic description and analysis of five different experiences that have the characteristics of living labs. The findings show that tourism living labs offer students opportunities for hands-on engagement in the co-creation and testing of frontier solutions with private, public and civil society sector partners. They also enhance social inclusion, environmental responsibility and life-long learning. For the tourism industry, labs can offer new knowledge; more, extended and deepened relationships; and opportunities to find an educated workforce. The challenges include project timeframes; documentation and information; equality among participants: the size of student groups; and the resources of the university. The article stresses the importance of recognising and handling the challenges that come with the introduction of living labs in a tourism context.

Introduction

Sustainable development and collaboration is the ideal of higher education today (Evans, Jones, Karvonen, Millard, & Wendler, 2015), not least in tourism programmes and courses (Coghlan, 2015). Higher educational institutions, or parts of them, can 'help society make the transition to sustainable life styles' (Velazquez, Munguia, Platt, & Taddei, 2006, p. 818) and function as integrated research and innovation arenas for co-creation processes (Zen, 2017). This calls for educational environments that inspire, challenge and develop students, and it requires an understanding of the complex relationships between education, research, collaboration and innovation (Gabrielsson, 2017). In tourism, several scholars point out that there is a gap between what the tourism industry expects from graduated students and what the academic institutions offer (e.g. Rodriguez-Anton, del Mar Alonso-Almeida, Andrada, & Pedroche, 2013; Wang, Ayres, & Huyton, 2010). The increasing complexity of tourism and the rapid changes in the market due to digitalisation and globalisation calls for new understandings of tourism as a phenomenon. It also requires cross-curricular skills in, for instance, problem-solving, information management, ethical behaviour and teamwork, which are difficult to teach (Rodriguez-Anton et al., 2013). Scholars request a balance in regards to content knowledge, vocational skills, liberal thinking and critical reflexivity in tourism education (Coghlan, 2015), which demands 'different approaches to learning, teaching and curriculum development' (Rodriguez-Anton et al., 2013, p. 34).

This paper focusses on co-creation, which distinguishes it from individual approaches such as internships and work placements. One of the concepts that appear in discussions about collaborative environments is 'living labs', typically categorised by open innovation, co-creation and experimentation in real-life settings (e.g. Gascó, 2017; Hawk, Romine, & Bartle, 2012). Living labs provide spaces to share ideas and develop, for example, goods, services, business models and systems, and if the lab is part of an academic institution, students get the opportunity to collaborate with stakeholders (Hawk et al., 2012) through active, effective and reflexive learning experiences (Zen, 2017). The purpose of this paper is to explore the concept of living labs in relation to education, innovation and sustainable tourism. A literature review is followed by five illustrations of living labs that the author has been involved in, concerning, for instance, how they are organised, time frames, what features they involve, how learning takes place and what critical matters come up. Finally, a conclusion with proposed contributions are given.

Innovation and sustainable tourism

The concept of sustainable tourism has developed and matured during the last three decades into an orientation aiming for environmentally, economically, culturally, socially and politically sustainable development through changes in behaviour and societal systems (Bramwell, Higham, Lane, & Miller, 2017). Organisations and people take responsible actions, which manifest in, for instance, CSR programmes in the tourism industry, through regulations, infrastructure and promotion from the part of public authorities, and through the individual and collective actions of tourists (e.g. Goldstein, Cialdini, & Griskevicius, 2008; Shove, 2010). However, as Bramwell et al. (2017, p. 7) put it, future directions must include undertaking 'complex issues in new ways, which demand a new order of collaborations that transcend disciplines and methodologies'. Knowledge transfer and accumulation of new knowledge are key for innovations to come forth (Milwood & Zach, 2016). Ideas for innovation are stimulated when you challenge your assumptions and adopt different perspectives on existing situations (De Bono, 1971; Moscardo, 2008). However, the tourism sector features small, seasonal dependent, competitive and fragmented businesses (Hjalager, 2010) with a shortage of resources, a high administrative burden and high rate of human resource renewal, which lower the absorptive capacity of the industry (Najda-Janoszka & Kopera, 2014). It results in an industry with a negative image on highly skilled human resources and leads to difficulties in attracting personnel and financial investments, which further weakens the industry's knowledge management culture and innovative capacity (Najda-Janoszka & Kopera, 2014, drawing on, for example, Cordeiro & Vieira, 2012; Nordin, 2003). Human resources are the key barrier to innovation in tourism, relating to, for instance, 'insufficient skills, competencies and low formal qualifications, as well as motivation to engage in innovation processes' (Najda-Janoszka & Kopera, 2014, p. 199). The difficulties in creating a collaborative innovation culture (McPhee, Guimont, & Lapointe, 2016) often means that new ideas are abandoned at a conceptual stage (Najda-Janoszka & Kopera, 2014). Maybe as a result, most innovations in tourism come forth with the influence of, or as a consequence of, other and wider innovations outside the tourism sector (Bramwell & Lane, 2012; Hjalager, 2015). Furthermore, since innovation in tourism can seldom be formally protected, 'free-riding' and imitation is common; any advantage you may gain will rapidly be wind-swept by competitors (Hjalager, 2010). Therefore, to develop sustainable competitive advantage, an important part of innovation in tourism also lies in the branding of places, destinations and products (Jernsand, 2016). The proposed potential lies in new methodologies (Hjalager, 2010), open processes (Sørensen & Sundbo, 2014) and the ability to look outside disciplinary boundaries to find developments crucial for innovation in tourism (Hjalager, 2010). Similarly, Najda-Janoszka and Kopera (2014, p. 199) propose the 'application of new media for education and knowledge transfer'.

Co-creation

Co-creation originates from the perspective that value cannot be delivered by firms, but derives in relational exchanges and through co-production (Gummesson & Grönroos, 2012; Vargo & Lusch, 2008). In Prahalad and Ramaswamy's seminal article from 2004, they particularly recognise and develop customers' (end-users') active roles as co-creators of value, although increasingly, customer communities, partners and employees are also included (Ramaswamy & Gouillart, 2010).

The types of values that derive from co-creation mainly refer to economic/monetary, social/cultural and/or symbolic/semiotic values (Graeber, 2001; Sanders & Simons, 2009). For instance, in the public sector, the involvement of citizens in public service delivery gives economic yields, such as effectiveness and efficiency, while other examples consider the act of citizen involvement itself as the main objective (Voorberg, Bekkers, & Tummers, 2015). Design scholars Sanders and Simons (2009) point out that '[o]ne of the key values of value co-creation is that it satisfies the need for creative activity while addressing the need for social interaction'.

In the literature on urban development and transdisciplinary research, the similar term co-production refers to inclusive processes that capture the understandings of different issues in particular contexts through the collaboration between a variety of actors (Gibbons et al., 1994; Polk & Kain, 2015). Knowledge is co-produced across disciplines (Guggenheim, 2006) and addresses 'the needs for democratic participation or more inclusive political processes called for within the sustainable development debate' (Polk & Kain, 2015, p. 33). Thus, the view on co-creation is wider, focusing on knowledge and understanding through interrelationships between academic, private and public actors and with a specific focus on sustainability, which is interesting in relation to the wide set of stakeholders and the holistic view that is needed for sustainable tourism. This paper understands co-creation as

> an inclusive and creative process between a variety of actors, where knowledge and ideas are captured,

integrated, generated and developed into sustainable innovations, and where the values deriving from the process are e.g. economic, social, cultural, environmental, experiential, symbolic, democratic and societal.

Living labs: co-creation and innovation arenas

Co-creation requires 'a common space where emotions, values, choices, ideas, and ideals emerge, converge or collide' (Campos, Mendes, Valle, & Scott, 2015, p. 21; drawing on Bochner, Cissna, & Garko, 1991). In Grönroos and Gummerus (2014, p. 209) description, 'a co-creation platform takes form when two or more actors' […] processes merge into one collaborative, dialogical process, in which the actors actively influence each other's processes and outcomes'. Such collaborative environment may have the form of a living lab, which this paper argues can be of use for innovation and learning across disciplines and stakeholders in tourism. The living lab is thus an arena for sustainable co-creation and innovation processes.

The concept of living labs was developed by the Smart Cities research group at Massachusetts Institute of Technology (MIT) (Hawk et al., 2012). The group integrated design and technology approaches to improve cities' resource efficiency and responsiveness to inhabitants needs and desires (MIT Media Lab, 2017). Instead of traditional closed and technology-based environments (e.g. R&D departments), living labs represent the concept of open innovation, being more user/human centred, inclusive and creative (Hawk et al., 2012). For instance, according to JPI Urban Europe (2015), an urban living lab is

> A forum for innovation, applied to the development of new products, systems, services, and processes in an urban area; employing working methods to integrate people into the entire development process as users and co-creators to explore, examine, experiment, test and evaluate new ideas, scenarios, processes, systems, concepts and creative solutions in complex and everyday contexts.

Living labs can be hosted by universities, public authorities or private sector organisations and they mainly act in a defined context such as a city or a region. Niitamo, Kulkki, Eriksson, and Hribernik (2006) point to the importance of neutral spaces for living labs as they reduce the risk that participants' are being hampered by institutional 'lock-in-effects', such as incorporated norms, cultures and working methods. The triple, quadruple or penta helix model for stakeholder involvement is often used, which emphasises the importance of collaboration between actors in the public and private sector, academia and civil society. An analysis of the living labs that are members of the global network *Open living labs* (Open living labs, 2017), shows that the descriptions of the labs often underscore users as participants and sources of innovation, sometimes in the forms of test panel participants (e.g. Evolaris Mobile Living Lab) and sometimes more deeply involved in service design (e.g. HumanTech LivingLab). The labs are referred to as physical locations but they can also be virtual (e.g. Images and Résaux-ImaginLab). Some of the members of the Open living labs are networks of labs in a region (e.g. Discovery Innovation Lab) and the structures are sometimes described as innovative ecosystems (e.g. Paris and Co Urban Lab), which can involve or be linked to business support such as incubators and seed funds (e.g. Krakow Living Lab). The living labs are frequently involved with projects in health care (e.g. life quality for elderly people or childcare) or urban development (e.g. smart cities) and they often include features of service design and digital technologies, mainly for public sector service development. Some labs focus on circular economy (e.g. Småland Region Living Lab) or other specific aspects of sustainability such as the reduction of carbon emissions (e.g. Adelaide living laboratories). Other living labs are specifically considered live training sites for industry and students. For instance, TAMK living lab in Finland is hosted by Tampere University of Applied Sciences and combine 'educational excellence' and 'practically oriented, user-driven research, development and innovation actions' (Open living labs, 2017, online resource).

A design approach, or 'design thinking', has gained attention in regards to innovation in recent years (Kimbell, 2011). It refers to a designerly way of knowing and thinking (Cross, 2007, p. 41), which could be of value for firms and policy makers (Kimbell, 2011). A method used by service designers is 'experience prototyping' (Buchenau & Suri, 2000; Martin & Hanington, 2012), which means that they put themselves in the shoes of the user and explore the experience in action (Jernsand & Kraff, 2015). Furthermore, participatory design, or co-design, is an acknowledged approach that applies collective creativity (Sanders & Simons, 2009) and uses stakeholders' practical knowledge and experience (Krippendorff & Butter, 2007) to co-create new products and services using e.g. visual tools (Jernsand, 2016; Jernsand, Kraff, & Mossberg, 2015). User-centered and participatory design is particularly interesting in a tourism context. Tourists continuously want novel experiences (Mossberg, 2007; Sundbo, 2009) and increasingly, consumers are involved, and want to be involved, in the production of their own experience (Alsos, Eide, & Madsen, 2014). This closeness between firms and end-users in tourism means that the knowledge from customers and employees drives innovation (Fuglsang, Sundbo, & Sørensen, 2011).

Students' experiential learning

An essential point of departure for co-creation is learning, which in this paper focus on student learning and education since higher educational institutions have crucial roles in the creation of knowledge (Velazquez et al., 2006). Related to living labs, Zen (2017) proposes that when students are involved in case studies where they face the opportunities and challenges of a real-world setting and collaborate with other stakeholders, they are involved with effective and reflexive learning. Active, participatory learning means emotional involvement or commitment, which the students achieve in interactive and supportive contexts where both the activities and the outcomes from them are meaningful (Jernsand, 2017; Mathieson, 2014; Terenzini, 1999). Learning theories emphasise the importance of reaching higher learning loops, i.e. higher levels of consciousness and awareness, which is not possible without conversation, feedback and critical reflection (e.g. Argyris & Schön, 1996; Flood & Romm, 1996). Using an experimental approach, the students are able to explore various perspectives, navigate in the social situation of working with others, reflect on their own learning and enrich the educational outcomes (Reid, 2015).

A variety of student learning experiences are characterised by engaged co-creation. Experiential learning (e.g. Kolb, 1984, 2015), problem-based learning (Barrows & Tamblyn, 1980), phenomenon-based learning (Østergaard, Dahlin, & Hugo, 2008), project-based learning, research-led teaching and multidisciplinary studio learning are some of the concepts that may fall into a continuum of experiential co-created learning theories, approaches, styles and methods. For instance, problem-based learning (PBL), or discovery learning, is an approach to instruction where tutors merely support, guide and challenge the learners in their inquiry and discovery process. PBL promotes the development of self-directed learning (SDL) and critical thinking skills. The similar approach phenomenon-based learning (Østergaard et al., 2008) emphasises context-specificity and future wanted situations.

An experiential approach goes in line with the change of emphasis based on the Bologna declaration for education, focusing on learning outcomes and competences through meeting the needs of the labour market (employability) and a transition from a teaching to a learning approach (Bologna working group on qualifications frameworks, 2005). From the university's point of view, the campus can thus become an experiential learning test-bed for solutions to the challenge of sustainability (Marcus, Coops, Ellis, & Robinson, 2015). For instance, Zen (2017) proposes that the way living labs expose 'the campus society to the environment' they lead to 'accelerating the integration of sustainability science in the academic society' (pp. 951–952). From the point of view of researchers involved in living labs, it combines operation, teaching and research (Robinson, Berkhout, & Campbell, 2011). Zen (2017, p. 951) suggests that for improved output, researchers should 'apply their tools, equipment and systems or approach'.

Critical issues of co-creation, experiential learning and living labs

David A Kolb, the scholar who introduced the idea of experiential learning to the scientific community, describes his critics' view of the concept as 'haphazard, unreliable, and misleading' (Kolb, 2015, p. xxi). A common view is that it stands in opposition to classroom and lecture learning (e.g. Keeton & Tate, 1978; in Kolb, 2015). Such a perspective, according to Kolb, downplays the importance of 'thinking, analysis, and academic knowledge', and reduces experiential learning to merely 'an educational technique' (Kolb, 2015, p. xviii). Rather, experiential learning integrates the traditional and experiential in 'a spirit of cooperative innovation' (Kolb, 2015, p. 5).

However, it should be noted that preparing for and facilitating living lab projects takes more time and other types efforts and resources than regular teaching (e.g. Evans et al., 2015), which university systems are generally not designed for. Further, there are gatekeeping academics who are not familiar with, or unwilling to identify projects and arrange partnerships between students and stakeholders outside academia (Evans et al., 2015).

A time consuming fact in living labs is that what one person sees as obvious is not necessarily evident for other people (e.g. Husserl, 1970). In collaborative, cross- and trans-disciplinary work, the knowledge and experience from each of the members of a group is difficult to grasp and consider, especially within the frame of a project (Westberg, Polk, & Frid, 2013). For instance, the different disciplines need to learn each other's vocabulary (Botero & Saad-Suulonen, 2013) and understand the challenges that face each of the actors in a specific context (Bason, 2013). Similarly, it is difficult to think and act at more than one structural level (micro to meta) at the same time and to move between them (zoom in and out). This process takes time and there is a risk that prescheduled time will be used for meetings and keeping up to date with what is going on in the project, while there is no time left for critical issues related to results (Westberg et al., 2013).

Since the activities in living labs are often practical, equal relationships, 'humbleness and modesty' are essential (Hawk et al., 2012, p. 228). Jernsand (2017, drawing from e.g. Lave & Wenger, 1991), proposes three aspects of engagement to enhance learning: embodied and situated learning, relationship-building, and acknowledging and sharing of power. Hawk et al. (2012) suggest that the higher education systems in the Nordic countries (mainly referring to Finland) are less competitive than structures in other parts of the world, specifically pointing towards the hierarchical divisions in the United States. Lecturers in the Nordic countries are mentors who listen and advise rather than set directions, which make living labs in these countries easier to handle. Participants from other cultures may find it hard to adjust to these 'flat leadership/followership structures' (Hawk et al., 2012, p. 228). Within tourism, Jamal, Taillon, and Dredge (2011, p. 136) point to the 'sustainability-oriented actions' that are critical to undertake, and for students to develop skills for, concerning environmental but also ethical issues, such as inclusion, exclusion, belonging and identity.

Experiences of living lab settings

In the following section, some examples of living labs that the author has been involved in are described in chronological order, to illustrate how student learning, innovation and sustainable tourism can be enhanced and what critical issues come up. The various approaches to student living labs are assembled in Table 1. Using an ethnographic approach, the empirical material mainly originates from research diaries and notes from participatory observations. The first three labs are cases of action research, where the author and her colleague were embedded in practical work with 1) a municipality, 2) a business incubator and 3) an ecotourism guide group. In these cases, the empirical material also includes interviews with participants involved in the projects. The empirical material from the last two cases are more shallow observations than the previous, concerning projects where students were involved in two completely different contexts: 4) the unstructured, unfamiliar yet deeply engaging Kenyan environment and 5) a research context with established structures and close mentor communication and instruction. The reason for including these last cases is as examples of shorter interventions, which still have impact in regards to sustainable tourism development and student learning.

Bollebygd master project and thesis

Business and Design is a master programme in collaboration between the business school and the design school

Table 1. Approaches to student living labs as experienced by author.

Approach to student living lab	Empirical example	Time frame	Size of student group	Stakeholder engagement	Main sponsors	Location	Main characteristics
Action-oriented master course/ thesis project	Bollebygd master project and thesis	1 year	Small	Master's students, municipality, idea-based sector, residents, local businesses	Academy, municipality	Small municipality in West Sweden	Multiple helix collaboration, evolutionary, relational, available project space
Creative arena with incubator	Brewhouse business incubator and arena	1 semester to continuous	Individual or small	Ex-students, start-ups, business coaches, creative industry businesses, academy (ongoing evaluation)	Regional and transnational authorities (EU)	Arena for cultural industries in Gothenburg, Sweden	Networks, peers, coaching, multifunctional arena
Action-oriented PhD project	Ecotourism PhD project	4 years	Small	PhD students (multiple disciplines), peers, int. research and knowledge centre, local businesses, residents	Academy and development cooperation agency	Fishing village outside Kisumu, Kenya	Community-based, transdisciplinary, evolutionary, transparent, relational
Significant part of master's course	Reality Studio master's course	7 weeks (part of course)	Medium-sized	Master's students, peers, local organisations, residents	Academy	3rd largest city in Kenya and its environments	Community-based, context-specific, reflexive, self-realizing
Minor part of PhD course	Marine aquaculture PhD course	1 week (part of course)	Medium-sized	PhD students, peers, experts in each discipline	Academy	Research stations in West Sweden	Research-based (experts involved)

at University of Gothenburg. The author was part of the programme between 2008 and 2010. During the third semester, student groups chose from various design briefs written by public and private organisations in the Gothenburg region. The author from business administration and a student colleague from design started working in a project with the municipality of Bollebygd. Together with representatives of the municipality, the students discussed and changed the design brief as to suit both parties, which made the students' engagement in the project deeper although still useful for the municipality. The students were not consultants who were supposed to do 'a job': the learning process was the central part of the work, and since there was no actual money involved, it enabled the process to be more open and experimental from the municipality's side, thus connected to an iterative and open-ended innovation spiral (e.g. Jernsand & Kraff, 2015). The work resulted in development of ideas from residents, NGOs, firms, officials and politicians and ended with an exhibition and a treasure hunt in the city centre, as well as a report on strategic design and place branding. Since the municipality is very small (appr. 8 000 inhabitants), the students came to know the senior civil servants and politicians very well as the project evolved. Gaining their trust, they also got the opportunity to do their thesis work in Bollebygd. The municipality arranged a space where the students could sit and work, continue and develop the exhibition and meet people on a daily basis, since the premises was a free shop space in the city centre. The students developed a brand platform together with stakeholders, in workshops, discussions and through visual tools as well as their situated learning about the place. Sustainability was at the core of the project. For instance, they found the life giving local water and the proud inhabitants as central for the Bollebygd place brand.

The location of a lab can be just as important for its success as the collaboration itself. Consequently, it may not always be that the space should be neutral for all, as Niitamo et al. (2006) suggest, or at least not the place, since tourism is place-specific. Living labs builds on real-life settings, and co-creation is at the core (Gascó, 2017; Hawk et al., 2012). Thus, it is essential to emphasise the specificity of the place along with the neutrality of the facilitators and the space. In the Bollebygd project, the fact that the author and her colleague had an available project space in the town centre made it possible for people to drop by. It was their town, and they were able to tell things just because they were there, but they did not have to face some local officials with whom they may have had a bad relationship. For instance, they could give and vote for ideas on the vision and core values of the municipality. The central location also made it possible for stakeholders to come who would not have been invited to other types of gatherings. From the part of the students, the situatedness was an embodied quality of learning, 'an evolving form of membership' in the community (Lave & Wenger, 1991, p. 53), which made them able to reflect on the situation of residents, officials, politicians, firms and tourists.

Brewhouse business incubator and arena

The engagement in student living labs can lead to new friendships and colleagues, new networks and new career opportunities. The author and her colleague found themselves working well together and started a company. One of the assignments was the ongoing evaluation of a business incubator for creative industries. The participants were often students with degrees from the university's arts, design, film and music programmes. The creative and permitting environment with studios, a large scene and supporting people from various creative sectors formed a living lab in itself. Students and other entrepreneurs started small businesses facilitated by the organised (and neutral) environment. In the interactions with the other participants in the same working space, they were able to reflect on their own situation, get continuous feedback and reach higher learning loops (Argyris & Schön, 1996). Further, the need for creative activity could be satisfied through social interaction (Sanders & Simons, 2009), which resulted in 'a spirit of cooperative innovation' (Kolb, 2015, p. 5) and a culture based on trust that is often missing in the tourism industry (McPhee et al., 2016). The setting of different types of companies also connects to that other sectors, outside tourism, influence the tourism industry; firms may even copy what is already out there and place it in their own context (Hjalager, 2010). Even though only a couple of the start-ups in the incubator were related to tourism, those that actually were could learn from the other participants.

Ecotourism PhD project

When the author and her colleague started their PhD studies, they continued working together in a transdisciplinary project in Kisumu, Kenya. A group of PhD students from Kenya and Sweden started collaborating with a tour guide organisation in a small fishing village by Lake Victoria and developed, for instance, their guided tours, a signage and waste management system and an association for male and female tour guides in the county. The author and her colleague explored and reflected on participation and inclusiveness in regards to sustainable tourism development, design and place branding.

Tourism experience innovation is a process where consumers play important roles as co-creators of their own experience (Alsos et al., 2014). In the living lab, the PhD students, the tourists and the members of the tour guide group were all involved in the innovation process, through an open process of planning, prototyping, testing and evaluation of a new type of guided tour (Jernsand et al., 2015). The group took advantage of the closeness between customers and employees (Fuglsang et al., 2011), the recognition of end-users (Ramaswamy & Gouillart, 2010) and combined it with academic knowledge (e.g. new methodologies) from multiple disciplines (Hjalager, 2010).

Including students in tourism experience innovation means that the entrance level for students is low since they probably have a relation to tourism and can put themselves in the shoes of the user (Buchenau & Suri, 2000; Martin & Hanington, 2012). This designerly way of working is often used in living labs (Open living labs, 2017) in which students learn new approaches and methodologies. In the Kenya project, the author and her colleague refined their competencies in design and marketing by using them in a new context, while other participants were new to the methodology. A while into the project, the guides declared that they had been asked by colleagues from other sites to teach them how to develop their site. A living lab can thereby be of use for spreading tools and methodologies, as Robinson et al. (2011) and Zen (2017) suggest. The tour guide association asked for prototypes, visual presentations and documentation, to be able to apply for funding for implementation from other beneficial organisations (see also Jernsand, 2016).

The PhD students met the participants' families and other residents, and got an understanding of the place, which would probably not have been possible through activities in the city centre of Kisumu. Also, the fact that workshops were held on the beach, right where the guided tours were to be held, made it possible to relate to the surroundings when composing guided tours. The examples points to situated learning and relationship building as vital parts of sustainable tourism development (Jernsand, 2017).

The comments and discussions with two female Kenyan PhD student colleagues made the author and her colleague aware of the importance of including women in tour guiding, which highlights the ethical considerations and critical thinking that students must undertake (Jamal et al., 2011). Tourism's complex nature with multiple stakeholders and environmental, socio-cultural and economic challenges related to the place makes sustainability at the core. The ecotourism setting exposed the author to these challenges: to involve people, protect the environment and improve the economic situation that many of the participants and residents are in. The types of knowledge and skills that students and others may learn also consider entrepreneurship and project management (Robinson et al., 2011), which are useful in tourism industry and governance.

The transdisciplinary research setting with specific emphasis on co-creation of value for people and society made the Kenyan living lab relevant as an example of sustainable tourism and innovation, as expressed by, for example, Bramwell et al. (2017). However, a vital factor for the 'sustainability-oriented actions' (Jamal et al., 2011, p. 136) was the length of the PhD project (four years). There is risk that projects with too short time frames and too much focus on results end up doing more harm than good, which the author realized the extent of, having worked in Kenya for some time. Coming from Sweden with frameworks, tools and expectations, the author and her colleague felt they set the agenda of the first parts of the project (see also Kraff, 2018; Kraff & Jernsand, 2014), which calls for ethical guidelines and preparations before entering living lab settings, especially in unfamiliar contexts.

Further, regarding the critical issue of time, it can be precarious when working in collaborative settings. Students from different disciplines and other stakeholders may have diverse timeframes and schedules, which complicates teamwork (Westberg et al., 2013). In the author's project in Kenya, a group of PhD students worked together, however, it was not without impediments. The Kenyan PhD students worked full time as teachers but when the Swedish team came to Kenya for three weeks at a time, they wanted to work intensively during a short period (see also Jernsand, 2016; Jernsand & Kraff, 2016). This often made the Swedish team take project leadership, which influenced the results in many ways. It calls for structures and facilitators for living labs, but it also calls for an understanding of the time factor (Jernsand, 2017). Furthermore, higher education systems varies in different parts of the world. The leadership structures with mentors, influenced by problem- or phenomenon-based learning rather than hierarchical divisions, make living labs easier to understand and handle for supervisors, lecturers and students from the parts of the world where equality is a central part of the political systems and culture (Hawk et al., 2012).

Reality Studio master course

During her PhD studies and after, the author has held lectures on participation, innovation and ecotourism and supervised student groups that are part of 'Reality Studio', a master course at Chalmers university of Technology in Gothenburg. During the course, the students spend about seven weeks in Kenya or Tanzania and work in practice-based projects related to design, architecture and urban planning. The experience of other cultures and societal systems give meaning to students' work. They get committed to the place, and often, they want to come back and work or become volunteers in similar contexts. Here, the context-specificity (Robinson et al., 2011) is one of the key learning aspects of the living lab. The real-life setting emphasises emotional involvement (Mathieson, 2014; Terenzini, 1999) and reflexivity (Zen, 2017).

The potential innovative results that come out of student living labs are of course one of the major raisons d'etre. If a lab can contribute to sustainable development in one way or another, it must be considered successful. In one of the Reality Studio projects in Kenya, the students designed a multilayer rack for drying fish on the beach, so that the fishmongers could get a better workplace and save space. Two years later when the author came back, the fishmongers had made a full-size rack themselves, which implies that seemingly small ideas can give effect through collaboration, documentation and information. It also implies that those students who are involved in living labs in interactive and supportive contexts find meaning in their work, confronting real challenges (Jernsand, 2017; Mathieson, 2014; Terenzini, 1999).

Marine aquaculture PhD course

The research centre Swemarc at University of Gothenburg organises a course that brings different disciplines together with a focus on marine aquaculture. The lecturers are, for instance, architects, marine biologists and social scientists. One of the lecture tracks is about sustainable maritime tourism, in which the author taught during the first course held in 2017. PhD students from all over the world gathered at Tjärnö and Kristineberg science stations in northern Bohuslän on the Swedish west coast. The science stations offer real-life working environments and the PhD students got one week together there, with lectures and group work. Thereafter, they continued working together in their home universities to fulfil the assignments on sustainable marine aquaculture. As Marcus et al. (2015) propose, the campuses at Tjärnö and Kristineberg constitute experiential learning test-beds: the students can better relate to the environmental issues since they are there, where the challenges of sustainability can be seen, reflected upon and handled. The type of living lab can also combine operation, teaching and research (Robinson et al., 2011), for instance by students' testing new tools, equipment, systems or analytical frameworks. However, it should be noted that the circumstances and conditions of a course often makes living labs hard to handle due to the number of students and the university's resources. Large student groups with few lecturers make the living lab approach even harder to accomplish.

Conclusion

Tourism's rapid growth (Hunt, 2017) and customers' demand for continuous novelty and surprise (Mossberg, 2007; Sundbo, 2009) must be set in relation to its environmental and socio-cultural impact (Budeanu, 2007). These and other challenges that tourism stakeholders face require new understandings, knowledge and skills that academic institutions of today are not sufficiently equipped to handle. A collaborative approach, innovative initiatives (Bramwell & Lane, 2012; Bramwell et al., 2017), new methodologies and multi-disciplinarity (Hjalager, 2010) are said to be the way to go to.

Student living labs make it possible to co-create and test frontier solutions with private, public and civil society sector partners. Fusing intellectual disciplines, labs can offer students opportunities for hands-on engagement in sustainable product development that aims for exciting tourism experiences, but also for social inclusion, environmental responsibility and life-long learning. For the tourism industry, the living lab is also an opportunity to find well-educated workforce with content knowledge and vocational skills, but also with abilities to reflect critically on tourism as phenomenon, system and practice. Thus, student living labs can function as innovation arenas for co-creation between stakeholders (Zen, 2017) aiming for a sustainable future for the education institutions and society (Velazquez et al., 2006).

The challenges with student living labs include e.g. timeframes, the size of student groups, the resources of the university, the interest from employees, information within and outside the organisation/lab, and equality among participants (also including non-participants). These challenges must be handled within the university system but also in each case, adapted to e.g. places, actors and seasons. A living lab in Kisumu, Kenya is different from one in Bollebygd, Sweden. There must be time and other resources to identify relevant projects, arrange partnerships before arranging a lab and also to handle critical issues within the lab. Recognising these impediments, the introduction of student living labs in a tourism context can be handled with consciousness. When critical thinking and experiential active learning go hand in hand, the values of co-creation lie both in the process and in its innovative results.

Disclosure statement

No potential conflict of interest was reported by the author.

References

Alsos, G. A., Eide, D., & Madsen, E. L. (2014). Introduction: Innovation in tourism industries. In G. A. Alsos, D. Eide, & E. L. Madsen (Eds.), *Handbook of research on innovation in tourism industries* (pp. 1–24). Cheltenham: Edward Elgar. doi:10.4337/9781782548416

Argyris, C., & Schön, D. A. (1996). *Organizational learning II: Theory, method and practice*. Reading, MA: Addison-Wesley.

Barrows, H. S., & Tamblyn, R. M. (1980). *Problem-based learning: An approach to medical education*. New York, NY: Springer Publishing Company.

Bason, C. (2013). Discovering co-production by design. In E. Manzini, & E. Staszowski (Eds.), *Public and collaborative: Exploring the intersection of design, social innovation and public policy* (pp. viii–vxix). DESIS Network. Retrieved from https://www.desisnetwork.org/wp-content/uploads/2017/04/DESIS_PUBLIColab-Book.pdf

Bochner, A. P., Cissna, K. N., & Garko, M. G. (1991). Optional metaphors for studying interaction. In B. M. Montgomery & S. Duck (Eds.), *Studying interpersonal interaction* (pp. 16–34). New York: Guilford.

Bologna working group on qualifications frameworks. (2005). *A framework for qualifications of the European higher education area*. Copenhagen: Ministry of Science, Technology and Innovation.

Botero, A., & Saad-Suulonen, J. (2013). Peer-production in public services: Emerging themes for design research and action. In E. Manzini, & E. Staszowski (Eds.), *Public and collaborative: Exploring the intersection of design, social innovation and public policy* (pp. 1–13). DESIS Network. Retrieved from https://www.desisnetwork.org/wp-content/uploads/2017/04/DESIS_PUBLIColab-Book.pdf

Bramwell, B., Higham, J., Lane, B., & Miller, G. (2017). Twenty-five years of sustainable tourism and the Journal of sustainable tourism: Looking back and moving forward. *Journal of Sustainable Tourism, 25*(1), 1–9.

Bramwell, B., & Lane, B. (2012). Towards innovation in sustainable tourism research? *Journal of Sustainable Tourism, 20*(1), 1–7.

Buchenau, M., & Suri, J. F. (2000). Experience prototyping. In D. Boyarski, & W. A. Kellogg (Eds.), *Proceedings of the 3rd conference on designing interactive systems: Processes, practices, methods, and techniques* (pp. 424–433). New York, NY: Association for Computing Machinery (ACM).

Budeanu, A. (2007). Sustainable tourist behaviour–a discussion of opportunities for change. *International Journal of Consumer Studies, 31*(5), 499–508.

Campos, A. C., Mendes, J., Valle, P. O. D., & Scott, N. (2015). Co-creation of tourist experiences: A literature review. *Current Issues in Tourism, 21*(4), 369–400.

Coghlan, A. (2015). Using scenario-based learning to teach tourism management at the master's level. *Journal of Hospitality and Tourism Education, 27*(1), 1–9.

Cordeiro, A. N. A. S., & Vieira, F. D. (2012). *Barriers to innovation in SMEs: an international comparison*. II Conferencia Internacional de Integracao do Design, Engenharia e Gestao para a inovacao Florianopolis, SC, Brasil. Brasil.

Cross, N. (2007). From a design science to a design discipline: Understanding designerly ways of knowing and thinking. In R. Michel (Ed.), *Design research now: Essays and selected projects* (pp. 41–54). Basel: Birkhäuser.

De Bono, E. (1971). *Lateral thinking for management: A handbook*. Harmondsworth: Penguin.

Evans, J., Jones, R., Karvonen, A., Millard, L., & Wendler, J. (2015). Living labs and co-production: University campuses as platforms for sustainability science. *Current Opinion in Environmental Sustainability, 16*, 1–6.

Flood, R. L., & Romm, N. R. A. (1996). *Diversity management: Triple loop learning*. Chichester, UK: Wiley.

Fuglsang, L., Sundbo, J., & Sørensen, F. (2011). Dynamics of experience service innovation: Innovation as a guided activity – results from a Danish survey. *The Service Industries Journal, 31*(5), 661–677.

Gabrielsson, J. (2017). Studenter bär på innovationskraften. *Entré, 3*, 24.

Gascó, M. (2017). Living labs: Implementing open innovation in the public sector. *Government Information Quarterly, 34*(1), 90–98.

Gibbons, M., Limoges, C., Nowotny, H., Schwartzman, S., Scott, P., & Trow, M. (1994). *The new production of knowledge: The dynamics of science and research in contemporary societies*. London: Sage.

Goldstein, N. J., Cialdini, R. B., & Griskevicius, V. (2008). A room with a viewpoint: Using social norms to motivate environmental conservation in hotels. *Journal of Consumer Research, 35*(3), 472–482.

Graeber, D. (2001). *Toward an anthropological theory of value: The false coin of our own dreams*. New York: Palgrave.

Grönroos, G., & Gummerus, J. (2014). The service revolution and its marketing implications: Service logic vs service-dominant logic. *Managing Service Quality, 24* (3), 206–229.

Guggenheim, M. (2006). Undisciplined research: The proceduralisation of quality control in transdisciplinary projects. *Science and Public Policy, 33*(6), 411–421.

Gummesson, E., & Grönroos, C. (2012). The emergence of the new service marketing: Nordic school perspectives. *Journal of Service Management, 23*(4), 479–497.

Hawk, N., Romine, M., & Bartle, G. (2012). The living labs: Innovation in real-life settings. *The Quarterly Review of Distance Education, 13*(4), 225–231.

Hjalager, A.-M. (2010). A review of innovation research in tourism. *Progress in Tourism Management, 31*(1), 1–12.

Hjalager, A.-M. (2015). 100 innovations that transformed tourism. *Journal of Travel Research, 54*(1), 3–21.

Hunt, E. (2017). Tourism kills neighbourhoods: how do we save cities from the city break? *The Guardian*. Retreived from https://www.theguardian.com/cities/2017/aug/04/tourism-kills-neighbourhoods-save-city-break

Husserl, E. (1970). *The crisis of European sciences and transcendental phenomenology: An introduction to phenomenological philosophy*. Evanston: Northwestern University Press.

Jamal, T., Taillon, J., & Dredge, D. (2011). Sustainable tourism pedagogy and academic-community collaboration: A progressive service-learning approach. *Tourism and Hospitality Research*, *11*(2), 133–147.

Jernsand, E. M. (2016). *Inclusive place branding: What it is and how to progress towards it* (PhD thesis), University of Gothenburg, Gothenburg.

Jernsand, E. M. (2017). Engagement as transformation: Learnings from a tourism development project in Dunga by Lake Victoria, Kenya. *Action Research*, *15*(1), 81–99.

Jernsand, E. M., & Kraff, H. (2015). Participatory place branding through design: The case of Dunga beach in Kisumu, Kenya. *Place Branding and Public Diplomacy*, *11*(3), 226–242. doi:10.1057/pb.2014.34

Jernsand, E. M., & Kraff, H. (2016). Collaborative PhDs: new approaches, challenges and opportunities. In H. Palmer & H. Walasek (Eds.), *Co-production in action: towards realizing just cities* (pp. 74–83). Gothenburg: Mistra Urban Futures.

Jernsand, E. M., Kraff, H., & Mossberg, L. (2015). Tourism experience innovation through design. *Scandinavian Journal of Tourism and Hospitality*, *15*(Sup 1), 98–119. doi:10.1080/15022250.2015.1062269

JPI Urban Europe. (2015). *Transition towards sustainable and liveable urban futures*. Retrieved from https://jpi-urbaneurope.eu/app/uploads/2016/05/JPI-Urban-Europe-SRIA-Strategic-Research-and-Innovation-Agenda.pdf

Keeton, M. T., & Tate, P. J. (eds.). (1978). *Learning by experience - what, why, how* (No. 1). San Francisco: Jossey-Bass.

Kimbell, L. (2011). Rethinking design thinking: Part I. *Design and Culture*, *3*(3), 285–306.

Kolb, D. A. (1984). *Experiential learning: Experience as the source of learning and development*. Englewood Cliffs, NJ: Prentice-Hall.

Kolb, D. A. (2015). *Experiential learning: Experience as the source of learning and development* (2nd ed.). Upper Saddle River, NJ: Pearson.

Kraff, H. (2018). *Exploring pitfalls of participation and ways towards just practices through a participatory design process in Kisumu, Kenya* (Doctoral thesis), Gothenburg University: ArtMonitor.

Kraff, H., & Jernsand, E. M. (2014). Designing for or designing with? In E. Bohemia, A. Rieple, J. Liedtka, & R. Cooper (Eds.), *Proceedings of the 19th DMI academic design management conference: Design management in an Era of Disruption* (pp. 1596–1611). Boston: Design Management Institute.

Krippendorff, K., & Butter, R. (2007). Semantics: Meanings and contexts of artifacts. In H. N. J. Schifferstein, & P. Hekkert (Eds.), *Product experience* (pp. 353–376). New York, NY: Elsevier.

Lave, J., & Wenger, E. (1991). *Situated learning: Legitimate peripheral participation*. Cambridge, UK: Cambridge University Press.

Marcus, J., Coops, N. C., Ellis, S., & Robinson, J. (2015). Embedding sustainability learning pathways across the university. *Current Opinion in Environmental Sustainability*, *16*, 7–13.

Martin, B., & Hanington, B. M. (2012). *Universal methods of design: 100 ways to research complex problems, develop innovative ideas, and design effective solutions*. Beverly, MA: Rockport Publishers.

Mathieson, S. (2014). Student learning. In H. Fry, S. Ketteridge, & S. Marshall (Eds.), *A Handbook for teaching and learning in higher education. Enhancing academic practice* (4th ed., pp. 63–79). New York, NY: Routledge.

McPhee, C., Guimont, D., & Lapointe, D. (2016). Editorial: Innovation in tourism. *Technology Innovation Management Review*, *6*(11), 3–5.

Milwood, P., & Zach, F. (2016). Innovative tourism destinations: Collaboration culture and absorptive capacity. *Tourism Travel and Research Association: Advancing Tourism Research Globally*, *16*. Retreived from http://scholarworks.umass.edu/ttra/2012/Oral/16

MIT Media Lab. (2017). *How buildings and cities can become more intelligently responsive to the needs and desires of their inhabitants*. Retrieved from https://www.media.mit.edu/groups/smart-cities/overview/.

Moscardo, G. (2008). Sustainable tourism innovation: Challenging basic assumptions. *Tourism and Hospitality Research*, *8*(1), 4–13.

Mossberg, L. (2007). A marketing approach to the tourist experience. *Scandinavian Journal of Hospitality and Tourism*, *7*(1), 59–74.

Najda-Janoszka, M., & Kopera, S. (2014). Exploring barriers to innovation in tourism industry–the case of southern region of Poland. *Procedia-Social and Behavioral Sciences*, *110*, 190–201.

Niitamo, V. P., Kulkki, S., Eriksson, M., & Hribernik, K. A. (2006). *State-of-the-art and good practice in the field of living labs*. Proceedings of the 12th International Conference on concurrent enterprising: Innovative Products and Services through Collaborative Networks (pp. 26–28). Italy: Milan.

Nordin, S. (2003). *Tourism clustering & innovation: Paths to economic growth & development*. Östersund: Etour.

Open living labs. (2017). *Living labs*. Retrieved from http://www.openlivinglabs.eu/livinglabs

Østergaard, E., Dahlin, B., & Hugo, A. (2008). Doing phenomenology in science education. A research review. *Studies in Science Education*, *44*(2), 93–121.

Polk, M., & Kain, J. H. (2015). Co-producing knowledge for sustainable urban futures. In M. Polk (Ed.), *Co-producing knowledge for sustainable urban development: Joining forces for change* (pp. 30–74). Abingdon, Oxon: Routledge.

Ramaswamy, V., & Gouillart, F. J. (2010). *The power of co-creation: Build it with them to boost growth, productivity, and profits*. New York: Free Press.

Reid, S. R. M. (2015). Tourism and design: Participatory inquiry as a possible route to innovation in tourism. *Tourism Dimensions*, *2*(2), 49–53.

Robinson, J., Berkhout, T., & Campbell, A. (2011). *The university as an agent of change for sustainability. Policy at a glance*. Toronto: Government of Canada: Policy Horizons Canada.

Rodriguez-Anton, J. M., del Mar Alonso-Almeida, M., Andrada, L. R., & Pedroche, M. C. (2013). Are university tourism programmes preparing the professionals the tourist industry needs? A longitudinal study. *Journal of Hospitality, Leisure, Sport & Tourism Education*, *12*(1), 25–35.

Sanders, L., & Simons, G. (2009). A social vision for value co-creation in design. *Open Source Business Resource*. Retreived from http://timreview.ca/article/310.

Shove, E. (2010). Beyond the ABC: Climate change policy and theories of social change. *Environment and Planning A, 42* (6), 1273–1285.

Sørensen, F., & Sundbo, J. (2014). Potentials for user-based innovation in tourism: The example of GPS tracking of attraction visitors. In G. A. Alsos, D. Eide, & E. L. Madsen (Eds.), *Handbook of research on innovation in tourism industries* (pp. 132–153). Cheltenham: Edward Elgar.

Sundbo, J. (2009). Innovation in the experience economy: A taxanomy of innovation organisations. *The Service Industries Journal, 29*(4), 431–455.

Terenzini, P. T. (1999). Research and practice in undergraduate education: And never the twain shall meet? *Higher Education, 38*, 33–48.

Vargo, S. L., & Lusch, R. F. (2008). Service-dominant logic: Continuing the evolution. *Journal of the Academy of Marketing Science, 36*(1), 1–10.

Velazquez, L., Munguia, N., Platt, A., & Taddei, J. (2006). Sustainable university: What can be the matter? *Journal of Cleaner Production, 14*, 810–819.

Voorberg, W. H., Bekkers, V. J., & Tummers, L. G. (2015). A systematic review of co-creation and co-production: Embarking on the social innovation journey. *Public Management Review, 17*(9), 1333–1357.

Wang, J., Ayres, H., & Huyton, J. (2010). Is tourism education meeting the needs of the tourism industry? An Australian case study. *Journal of Hospitality & Tourism Education, 22*(1), 8–14.

Westberg, L., Polk, M., & Frid, A. (2013). *Gemensam kunskapsproduktion för urbana förändringar (Joint knowledge production for urban change), version 1*. Gothenburg: Mistra Urban Futures.

Zen, I. S. (2017). Exploring the living learning laboratory: An approach to strengthen campus sustainability initiatives using sustainability science approach. *International Journal of Sustainability in Higher Education, 18*(6), 939–955.

'Dig where you stand': values-based co-creation through improvisation

José-Carlos García-Rosell, Minni Haanpää and Jenny Janhunen

ABSTRACT

The concept of co-creation has been extensively discussed in tourism and hospitality research. Most studies focus mainly on exploring the involvement of customers in experience creation and the development of tourism services and innovations. Although this way of thinking can be valuable for creating competitive tourism services, it neglects other aspects of co-creation that may play an important role in tourism. Considering this gap, we work towards a more comprehensive understanding of value co-creation processes in small tourism firms by drawing upon cultural marketing and organisational improvisation. To illustrate this process, we present two company cases from northern Sweden, Treehotel and Icehotel. The empirical material used in this study consists of qualitative data collected via semi-structured interviews, participant and non-participant observation, and documentary materials. The study contributes to the co-creation discussion in tourism and hospitality by unpacking the role of human and non-human stakeholders, material resources, and unexpected events in the value co-creation processes within a small tourism firm context.

Introduction

The concept of co-creation has been extensively discussed in the marketing literature over the last decades (e.g. Payne, Storbacka, & Frow, 2008; Peñaloza & Venkatesh, 2006; Prahalad & Ramaswamy, 2004; Vargo & Lusch, 2004). Much of the discussion has focused on the process of value creation and the role of customer knowledge, skills, and competencies as key resources in it (Prahalad & Ramaswamy, 2004; Vargo & Lusch, 2004). Indeed, when elaborating on the notion of value creation, Vargo and Lusch (2004) argue that value is produced in a collaborative and iterative learning process on the part of the firm and the customer. As Prahalad and Ramaswamy (2000) note, the market offers a venue for proactive customer involvement in the value creation process. According to this view, consumers become co-creators who use their skills and knowledge to create the offerings best suited to their needs (Vargo & Lusch, 2004).

This understanding of co-creation has also found fertile ground in tourism management studies. Tourism management scholars have explored co-creation as a strategic tool for actively engaging customers throughout the process of service development, experience creation, and innovation (e.g. Campos, Mendes, do Valle, & Scott, 2018; Hjalager & Nordin, 2011; Roeffen & Scholl-Grissemann, 2016). In doing so, they portray co-creation as a rational and systematic process controlled by the firm. Although the tourism management literature of co-creation offers a useful road map for designing and managing competitive tourism services (Campos et al., 2018), previous researchers fail to recognise other aspects of co-creation that may play an essential role in a small tourism firm context. For example, little attention has been given to the role of stakeholder relationships in co-creation and to the fact that the co-creation of value in small tourism firms is not necessarily a separate activity detached from the place and everyday working practices (see García-Rosell, Haanpää, Kylänen, & Markuksela, 2007; García-Rosell, Kylänen, Pitkänen, Tekoniemi-Selkälä, & Vanhala, 2015; Haanpää, García-Rosell, & Tuulentie, 2016).

Although small tourism firms have strong ties to their places, they are likely to lack financial resources, strategic planning, and management expertise (Ateljevic & Doorne, 2000; Schilar & Keskitalo, 2018; Valtonen, 2009; Yachin, 2019), which may constrain the implementation of co-creation processes, as suggested in the tourism management literature (Campos et al., 2018; Hjalager & Nordin, 2011). Rather than being concerned with these limitations, we view the characteristics of small tourism firms and their relations to their places and stakeholders as an opportunity to elaborate further on the notion of co-creation in tourism and hospitality studies. For instance, the smallness of tourism firms and their relational natures offer a fruitful ground for studying the roles of different stakeholders and multiple ways of knowing in value co-creation in tourism.

Drawing upon a cultural marketing approach to co-creation (Peñaloza & Venkatesh, 2006) and organisational improvisation (Kamoche & Cunha, 2001; Vera & Crossan, 2004), we explore the role of different stakeholders and the knowledge and skills that they mobilise, produce, and reproduce in value co-creation in tourism. More precisely, our study aims to illuminate how the notion of value co-creation is intrinsically related to the values and narratives of entrepreneurs and the full range of stakeholders who are part of the co-creation process. To that end, we draw on the case of two Nordic tourism firms in particular, Treehotel and Icehotel. This article is structured as follows. We start with a review of research on co-creation and organisational improvisation. Then we elaborate on the process of data collection and analysis. This is followed by a report on our study results – firm stories as told by the owners of the firms – and a discussion on the findings. We conclude with implications and suggestions for further research.

Value co-creation through stakeholder interaction and improvisation

As a business strategy, co-creation implies that firms focus on the processes that create value and maximise consumer participation in the customisation of services by means of using consumers' knowledge and skills (Vargo & Lusch, 2004). Although this idea of co-creation has been widely accepted in marketing and tourism studies in particular (Prebensen, Chen, & Uysal, 2014), the concept has been under continuous evaluation and development over the past decades. In particular, criticism has been directed towards the vagueness of the notion of value and the strong focus on firm–customer interactions and relationships.

Service marketing scholars have particularly been interested in elaborating further on the notion of value and how it is created (Grönroos & Voima, 2013; Vargo & Lusch, 2008). Grönroos (2012) suggested, for example, that instead of treating value metaphorically as originally done by Vargo and Lusch (2004), a more interactional, contextual treatment of the concept in direct customer–firm relations would offer better theorising. To further understand interaction and context in co-creation, marketing scholars have moved towards social constructionist, and more recently also practice-based, approaches (see e.g. Edvardsson, Tronvoll, & Gruber, 2011; Helkkula, Dube, & Arnould, 2018; Rihova, Buhalis, Moital, & Gouthro, 2013; Storbacka, Frow, Nenonen, & Payne, 2012). Co-destruction of value, instead of a successful co-creation process, has also received attention (Plé & Chumpitaz Cáceres, 2010; see also Echeverri & Skålén, 2011).

There is also a stream of cultural marketing literature which offers a critical evaluation of co-creation and thus alternative ways of theorising it (e.g. Cova, Dalli, & Zwick, 2011; Peñaloza & Venkatesh, 2006). Peñaloza and Venkatesh (2006) were among the first scholars to draw attention to consumers' subjective understandings and practices regarding their participation in the co-creation of value and power relations between consumers and firms. According to them, treating consumers as subjects and agents in the marketplace is a step forward to a more just treatment of the different actors involved in a co-creation process. Similarly, other studies have focused on the political aspects of co-creation and the role of consumer communities and practices in the co-creation of value (e.g. Cova et al., 2011; Echeverri & Skålén, 2011; Pongsakornrungsilp & Schroeder, 2011). In these studies, co-creation is conceptualised as 'a dynamic process in which multiple objects, actions and agencies interact' (Lugosi, 2014, p. 177).

For example, Echeverri and Skålén (2011, p. 353) conceptualise value as 'co-created, realised, and assessed in the social context of the simultaneous production and consumption process'. According to them, value cannot be measured in monetary terms; rather it is subjectively defined by the customer and the provider. As Holbrook (2006) points out, value is collectively produced in actions and interactions, but subjectively experienced. Value is symbolic, historically and culturally situated, and strongly connected to the life projects and narratives of consumers (e.g. Holbrook, 1999; Peñaloza & Venkatesh, 2006, p. 310; Pongsakornrungsilp & Schroeder, 2011, pp. 305, 309).

In line with the cultural marketing approach that we take in this study, we view value as a social construct that is created, maintained, negotiated, resisted, and transformed throughout stakeholder actions, interactions, and relationships (e.g. Holbrook, 1999; Moisander & Valtonen, 2006; Peñaloza & Mish, 2011; Peñaloza & Venkatesh, 2006). By elaborating on the nature and process of 'market co-creation', Peñaloza and Mish (2011) broaden the scope of value co-creation beyond consumers to include other market actors such as suppliers, investors, competitors, citizens, regulators, and even the natural environment. This view of co-creation is supported by recent tourism studies that have recognised the role of local residents, governmental organisations, and non-human actors in the co-creation of tourism services (e.g. Bertella, 2014; Haanpää et al., 2016; Lugosi, 2018; Rantala & Mäkinen, 2018).

Considering this, we argue that, in order to understand the process of co-creation in tourism, we need to explore the complex community dynamics which drive stakeholder interactions and thus contribute to the creation of different types of knowledge (see Dredge & Jenkins, 2011). Consistent with Valtonen (2009), we regard knowledge as something embedded in stakeholder relationships and continuously developed through the sharing of experiences, stories, tools, collaborative practices, and ways of doing things. These relationships and interactions are epistemic in nature, as they provide opportunities for tourism firms to acquire and generate knowledge, by observing, listening, discussing, reflecting, and acting together with their stakeholders (see Valtonen, 2009).

To grasp the dynamics of knowledge in a co-creation process, we draw upon an improvisation approach. Within organisation studies, there is a research tradition devoted to the theorisation of improvisation – the emergent and unexpected – as an essential element of organisational life (see Cunha, Neves, Clegg, & Rego, 2015). These studies have borrowed from the realms of improvisational theatre and jazz improvisation, where ideas are created and implemented in action (Crossan, Cunha, Vera, & Cunha, 2005). Under the term 'organisational improvisation', improvisation has been widely discussed in relation to research areas such as organisational learning (Barrett, 1998; Cunha et al., 2015), strategy (Crossan & Hurst, 2006; Eisenhardt, 1997), and service development (Kamoche & Cunha, 2001; Kyriakopoulos, 2011; Miner, Bassoff, & Moorman, 2001).

Although different definitions have been suggested to capture the nature of improvisation in an organisational context, all definitions converge on the idea that all forms of improvisation have a purpose, are extemporaneous, and occur in action (Cunha, da Cunha, & Kamoche, 1999). In this study, we follow Vera and Crossan (2004, p. 733) to define improvisation as the spontaneous and creative process of attempting to achieve an objective in a new way. Whereas 'spontaneous' refers to a flexible and unplanned process, 'creative' indicates a process aiming to develop something new and useful to the situation.

Several management and organisational scholars have shown that improvisation can be useful in understanding the process through which products and services are developed (e.g. Kamoche & Cunha, 2001; Kyriakopoulos, 2011; Miner et al., 2001). From this perspective, the idea of improvisation as collective actions that are not restricted to one-time events (Vera & Crossan, 2004) seems to be appropriate to explore the co-creation of value within a small tourism firm context. In line with Brown and Eisenhardt (1998, p. 33), we argue that improvisation is what nurtures the acquisition and generation of knowledge, enabling tourism firms to continuously co-create value in close collaboration with their stakeholders. Improvisation can be regarded as a form of real-time learning that arises when firms and their stakeholders struggle to deal with a problem or unexpected opportunity (see Miner et al., 2001, p. 306). According to Miner et al. (2001), improvisation can influence long-term trial-and-error learning, leading to new activities, insights, and ways of knowing.

Similar to the co-creation of value, improvisation also draws upon material and social resources that are available within a firm's operating environment (Cunha et al., 1999, p. 302). Whereas material resources encompass resources such as buildings, nature, landscapes, and their meanings, social resources refer to the social structures and relationships among stakeholders performing improvisation. In addition, improvisation draws from temporality: interpreting the present through past experiences, and anticipating the forthcoming in the operating environment and beyond (see Haanpää, 2017; Haanpää et al., 2016; Haanpää, García-Rosell, & Kyyrä, 2013). While organisational improvisation contributes to a better understanding of co-creation in a small tourism firm context, it should be noted that it is not always tied to success (see Kamoche & Cunha, 2001). Improvisation requires a good understanding of the various relations of the context; otherwise, it can also co-destruct value (see Echeverri & Skålén, 2011). In line with Vera and Crossan (2004), it is important to differentiate between how improvisation occurs in these processes and what it takes to do it well. In this study, we focus on two small tourism firms that help illustrate the role of improvisation in co-creation processes leading to value creation and innovative tourism products.

Understanding small tourism firms through narratives and observation

The study was conducted within the scope of two EU-funded development projects dealing with service development and co-creation in tourism. By taking an ethnographic approach, in the first project (2008–2011), we conducted participant/non-participant observations and attended approximately 20–25 service development meetings in seven tourism companies operating in Northern Finland over a period of two years. The data were complemented by 50 interviews with tourism companies operating in the same geographical area (García-Rosell et al., 2015; García-Rosell & Haanpää, 2017).

During the data collection stage of this first project, research team members met regularly to examine and reflect on the transcripts of interviews and field notes to identify common co-creation patterns of service development and co-creation. The data revealed that service development and co-creation processes were intertwined with the daily practices of tourism firms, their customers, and other stakeholders. Challenges or unexpected situations seemed to play a role in reaching breaking points, leading to new services and working practices. The common pattern of co-creation reflected in the data showed an ongoing, spontaneous, and goal-oriented learning process with no fixed structure.

In the second project (2013–2015), we continued our study by conducting further participant/non-participant observations and interviews, but this time with companies situated in northern Sweden. Data collection in the second project lasted for one year. Similar co-creation patterns as in Finland were identified in the Swedish companies involved in the study. During both projects, we informed the participating firms and their employees that we were conducting a study on service development. To avoid any biases, we did not use the word 'co-creation' in the projects and our study in particular. Our aim in both projects was to use different forms of ethnographic data to explore co-creation in relation to service development in small tourism firms.

During the process of data collection in the second project, we decided to put additional emphasis on two Swedish companies: Treehotel and Icehotel. Both companies are situated in the Swedish county of Norrbotten, which is by far the least densely populated region in Sweden. Indeed, the county includes less than 3% of the Swedish population yet it covers approximately 22% of Sweden's total area. The region is highly dependent on income from tourism and nature-based industries. We chose these two companies because the co-creation patterns identified in our longitudinal study are reflected throughout their history. These companies were also interesting cases due to their role in driving the economic and social development of their home villages and contribution to the brand 'Swedish Lapland' in the international market. Indeed, these two companies have been internationally acknowledged as highly innovative and unique in terms of the production of tourism experiences. It should be noted that prior to this study, two of the authors were familiar with these two companies through several on-site visits and cooperation activities organised within other EU projects.

The empirical material collected from these companies consisted of semi-structured interviews, participant and non-participant observation, and documentary materials. A team of four researchers participated in the collection of data. Together with the project manager, two of the authors were involved in data collection. The remaining author was a researcher participating in the project. In the first step, the team of researchers gathered documentary material (brochures, newspaper articles, blog posts, and videos available online) related to the two firms. These data were used to gain a better understanding of the firms' background, evolution, and services.

Documentary material was also needed to define the focus of the observation and interviews conducted in a later stage of the study (see Atkinson, 2002). Participant observations were conducted prior to each interview by two researchers who took part in a guided tour around the premises of the firms. Non-participant observations were also conducted in the premises of both firms by one researcher, while the second researcher was conducting the interviews. By writing field notes, the researchers not only kept a record of their observations but also reflected on their personal experiences and impressions during the firm visit (see Hammersley & Atkinson, 1996). Participant and non-participant observations were seen as an important part of the data collection phase because they enabled the researchers to look into the sociocultural and material context in which the daily organisational practices of the firms unfold (Jorgensen, 1989). Visits to each firm lasted between three and five hours including the interviews.

Semi-structured interviews were used as a means to gain a deeper understanding of the range of roles and organisational practices that exist within these two tourism firms (see Atkinson, 2002; Eriksson & Kovalainen, 2008, pp. 82–83). The interviews took place in the premises of the firms and were conducted with one of the co-owners of each hotel. The two persons interviewed were women and aged 50–55. The interviewees gave full consent to be recognisable and to use their real names within the context of this study. These semi-structured interviews lasted approximately 60 min, were recorded on audiotape, and then carefully transcribed. The semi-structured interviews followed the principles of life story interviewing due to the strong connection between the entrepreneurs' lives and their firms (see Bredvold & Skålen, 2016). We followed the constructionist life story approach, where the story is created in the constructive collaboration between the parties involved in the interview. There were some themes that were purposely discussed with the interviewees, but basically interviewees had the opportunity to construct the story of the firm and its relation to their life in their own terms (Atkinson, 2002). In line with the data collection, no reference to co-creation was made in the interviews. It is through their life stories in their firms that we could explore co-creation in relation to their services and business operations (see Atkinson, 2002).

Transcripts from interviews, field notes, and data collected online were thematically analysed. Both ethnographic and content analyses were used when interpreting the collected data (Rantala, 2011). By iterating between the data and the theoretical framework, we focus on understanding the process of co-creation as it happens in the two chosen companies. In doing so, we pay particular attention to identifying and illustrating stakeholders, material resources, and events in the processes of value co-creation.

The firm stories

Treehotel, Britta

Britta is one of the owners of Treehotel that is located in the small village of Harads (500 inhabitants) close to the Arctic Circle. The history of Treehotel goes back to the establishment of Brittas Guesthouse in 2004 – a guest house offering a homey atmosphere from the 1930s to 1950s and views of the river and forest. After six years of guest-house operations, Treehotel was established as a parallel company. Treehotel was inspired by the surroundings of Brittas Guesthouse and 'The Treelover', a documentary filmed in Harads. The documentary tells a story about a man who builds a traditional tree house to reminiscence about his childhood summers. The core idea of the Treehotel is to provide high-quality, unique accommodation service with an emphasis on sustainability, Scandinavian design, and nature. Currently, Treehotel is owned by five owners and employs five to eight people. It has seven 'tree rooms', one conference room, and two saunas in the forest.

Britta and her husband Kent were born in Harads. She worked in health care for nearly 20 years, and after that she spent 12 years working with her husband in rural development projects in Harads. One of the initial development projects she was involved in, took place in 2000 and was related to a listed house. The building used to be a retirement home more than three decades ago, and it had remained empty since then. In 2004, Britta made a decision to buy the house and turn it into a small guest house with a restaurant and services for conferences and private events. Although the operations of the guest house demanded a lot of work, it did not generate much profit. In 2006, the documentary 'The Treelover' was filmed by a friend of hers in Harads close to the guest house. Britta and Kent were so fascinated with 'The Treelover' that they started to rent the tree house built for the documentary. At that point, the tourism services they offered consisted of the guest house and the tree house. Nevertheless, new services needed to be developed in order to improve the profitability of the company.

During this time Kent was working as a school counsellor and in his own fishing company which offers tours around the world. The fishing company had a key role in the development of the Treehotel concept. It was during a fishing trip that Kent introduced the Treehotel idea to a group of architects, who had been taking part in the fishing tours once or twice a year for the last 10 years.

> They were on a fishing tour, 2008, and you know we rent out this tree house, the first tree house. And Kent, my husband said, 'I will take a picture and ask the guys [the architects] if they, or what they think about the idea to make these hotel rooms. So, he just asked them when they were sitting in front of the fire one evening with food and vodka. . . . They were good friends, we have known them for many years.

The architects found the idea fascinating, and soon one of the architects was in Harads with Britta and Kent conducting the pre-study and making the first drawings. Since then the couple has been contacted by approximately 30 architects around the world who would like to realise their designs in Treehotel. This wave of new ideas was triggered by the media, which put Treehotel in the spotlight both nationally and internationally.

> From the first year it was very difficult to see what was so interesting. We could not see it. We did not understand why BBC or CNN wanted to come here. . . . And why 4 million people wanted to listen to us on the radio.

The municipality and the county council played an essential role in facilitating the process and helping the company with all practicalities and formalities needed to start the project. They even arranged a meeting between the bank and the couple to get a loan for the realisation of the concept.

> We had a lot of help from the municipality. They helped us with everything. Because when we had this idea, and started talking with them in the municipality and the county council, we made a pre-study and we got 50% from the county council to make this pre-study. . . . They were here a lot and we made some plans how it's going to be and if it was possible to do. And then the municipality helped us a lot when we said it was ready and we must do this, because we didn't know but only had a gut feeling. It is such a good idea. And you know, our municipality helped us; we had never arranged a meeting, never.

In order to get the loan, Britta and Kent needed to own the land where the Treehotel was to be built. The landowner was a friend of theirs who just happened to have sold the land to a forestry company, which planned to cut down the trees. Nevertheless, he was so fascinated with the idea that he changed his mind and decided to terminate the sales agreement and sell the land to Britta and Kent. Since they did not have the money for the purchase, the landowner joined the company with the land, thus becoming one of the owners of Treehotel. Also the strong relationship Britta and Kent had with the place and the forest played an important role in developing the world-famous Treehotel concept.

> I said it must be something that is around this area, must be something pure. . . . It has to be special and different, but I mean you have to 'dig where you stand', it must be here. It must be real from this area. And what we do have here is forest. We have no mountains, but we have forest and we have not done anything in the forest.

The Treehotel has become a famous tourism attraction due to its uniqueness and the continuous positive international media coverage. Nowadays, Treehotel attracts thousands of visitors every year. While some visitors may choose to stay overnight in one of the seven tree rooms, others may just satisfy their curiosity with a tour around the tree rooms and the surrounding forest. Although there is a high demand for the tree rooms, and many requests from architects willing to realise their designs in Treehotel, the owners have made the decision to put a limit to the number of tree rooms. The last tree room, which was named Seventh Room, was built in 2016.

Icehotel, Kerstin

Kerstin is one of the owners of the first-ever 'ice hotel' in the world. The Icehotel is located about 200 kilometres above the Arctic Circle in Jukkasjärvi, a village in the municipality of Kiruna, northern Sweden. The story of the hotel can be traced back to the 1940s when the local community started developing services in the village. Today, the Icehotel offers accommodation in unique rooms made of ice designed by artists coming from all over the world. During the first years of the company, until 1990, it was partly owned by a local cooperative. In 2014, the Icehotel was run by five private owners and employed 20 people around the year and 135 during the high season. The Torne River has had a significant meaning for the firm and its development throughout its history.

> And this [Torne River] has been 'the mother of our company' from the beginning and still is. It's the Torne River. The Torne River has fed me for 28 years. And not only me. We are 150 people working in this company. . . . It gives us food and fresh water, fun, activities, art, good friends, and it just gives us so much. The water is everything for the company.

The first CEO of the company was Yngve Bergqvist. At the beginning the hotel was mainly operating during the summer season, offering activities and accommodation. The first activities of the company were canoeing and river rafting. In 1986, there was a serious river rafting accident in the area, and the activities dropped suddenly. After this incident, the owners of Icehotel had to rethink the business concept of the company.

> At the time, the summer was the peak season for us. . . . We were overbooked and almost everybody was doing the river rafting. But we did not have any guests in the winter. We tried to have people coming up in the winter but they said, 'Oh no Kerstin, it's too dark and it's too cold'.

In 1988 Bergqvist got a chance to travel all over the world. He chose to travel to winter destinations to look for business ideas for the long winter in northern Sweden. In Japan, Bergqvist visited a winter ice-sculpting festival where some of the artists had built the church of Jukkasjärvi from ice to honour his visit. Due to the festival, he became familiar with ice as a material and got the idea of using the Torne River to provide the ice for making the ice sculptures. Immediately the next winter, in 1989, Bergqvist invited two Japanese artists to Jukkasjärvi to teach the people in his company how to work with materials such as ice and snow. The first building made out of ice and snow was built in 1990. That was the beginning of the Icehotel as it is known today.

> In the spring of 1990, we built the first building of ice and snow. We did it on the ice down the Torne River. And that was a big mistake because when the spring was coming it would get very heavy and we would get water on the floor. So that is where everything started.

At that time, the current five owners of Icehotel were employees of the company, working in different roles and developing their expertise in tourism. This is something that they see as one of the cornerstones of the development of Icehotel. Also, close cooperation with the local community and entrepreneurs has been important for the owners. In relation to this, there was one specific event that shifted the development towards building a hotel from ice. A friend of Kerstin called her one day and wanted to come to Jukkasjärvi with a group to organise a conference there. Unfortunately, all cabins were booked at the time of the request, and Kerstin was not able to accommodate the group.

> I said, 'You cannot come. We are fully booked.' They said, 'We can sleep in a snow building'. I said, 'No, you can't sleep in a snow building. It is too cold.' 'Yes', they said. [Kerstin finally gave up saying] 'OK, you can sleep there. I will arrange some sleeping bags for you and reindeer hides. We even will make you a diploma for sleeping outdoors during winter.'

Eventually, the guests slept in the snow building and even got a diploma from it. They not only embraced the experience, they also became the first guests of the Icehotel. From that point on, the concept was developed further, learning by doing and from mistakes made, as Kerstin admitted.

> In March we start taking the ice from the Torne River in big blocks. . . . We store them in the house until next winter. We take up about 5000 tons of ice from the Torne River. We actually only borrow it because in April, May, and June everything is melting back to the river again. Today we do it with tractors when before we did it by hand. And we did it in December when the ice was not so thick. . . . When we started we couldn't go anywhere to ask how to do it. . . . At that time no one could help because no one had done it before. So every year we made a lot of mistakes, but every year we also developed the knowledge.

The core idea of the first-ever Icehotel is to offer meaningful, inspiring, and unique art, nature, culture, living, and food experiences. Today, building the Icehotel is grounded on a variety of stakeholders working tightly together. Every year the company looks for new artists to plan, design, and build unique hotel rooms from ice and snow. Artists from all over the world send their hotel room design ideas every year. From all ideas, the company representatives choose 25 who are then invited to Jukkasjärvi for two weeks during the winter to realise their designs in Icehotel. Also, other strong partnerships and cooperative relations have reinforced the development of the company in the tourism market.

The Icehotel is a unique story of developing a business concept based on the place, a chain of events, and people working with it. Working with the elements from the place – the river and ice and snow in particular – ended up defining the uniqueness of the business concept, that is still going strong today. The Icehotel is not only the main attraction of Jukkasjärvi but also one of the reasons for travelling to this part of Sweden. Icehotel has also enjoyed generous media coverage. It has also become one of the icons of Sweden as a Nordic tourism destination. In November 2016, Icehotel opened 'Icehotel 365', a permanent ice experience with an ice bar, gallery, and ice hotel rooms.

Discussion

From our analysis of Treehotel and Icehotel, we develop three insights relevant to the process of value co-creation through improvisation. First, we show how co-creation involves multiple human and non-human stakeholders. Second, we draw attention to the material resources upon which these firms and their stakeholders act to co-create value. Third, we point to challenges and unexpected opportunities, which trigger co-creation processes leading to innovative services.

Stakeholders

Our findings point to a multiple number of stakeholders who play an important role in the co-creation process. Both Treehotel and Icehotel acknowledge customers, employees, local authorities, local community members, artists, architects, the media, and other partners as actors who employ their knowledge and skills in service development and other joint activities (see Cabiddu, Lui, & Piccoli, 2013; Peñaloza & Mish, 2011; Valtonen, 2009). Our findings draw attention to the way these two tourism firms rely on their stakeholder relationships in dealing with challenging situations or issues they lack knowledge on. For example, the owners of Treehotel relied upon their long-time relations to a group of customers (the architects) to acquire and generate the knowledge and skills needed to put forward their tree house concept. Similarly, Icehotel collaborated from the very beginning with artists to generate the knowledge and skills needed to created buildings and structures made of snow and ice. In line with Peñaloza and Mish (2011), Treehotel and Icehotel do not view stakeholders as something to be managed, but rather as an extension of their organisations.

In our study, we identified fuzziness between stakeholders and material resources, which questions the prevailing logic separating human and non-human actors in the co-creation (Vargo & Lusch, 2004) and organisational improvisation literature (Cunha et al., 1999). Treehotel and Icehotel did not approach the natural environment just as a material resource to be used and enacted upon. Rather, consistent with Peñaloza and Mish (2011), these firms view the river and the forest as actors incorporating their agency in co-creation processes. The natural environment becomes a key stakeholder in the co-creation of value, enabling and limiting the actions of the firm and other stakeholders (see Lugosi, 2018). Treehotel and Icehotel have developed collaborative ways of knowing and being with nature that allow them to consider the interest of the forest and the river when developing their services and business practices (Rantala & Mäkinen, 2018). Indeed, there is a strong relationship between the firms and non-human actors, which are an intrinsic part of the places where they perform (see Schilar & Keskitalo, 2018).

Material resources

In considering the natural environment as a stakeholder, we discuss material resources here in terms of buildings, construction materials, and meanings that become an intrinsic part of the co-creation process. Consistent with Vargo and Lusch (2004) and Cunha et al. (1999), we see material resources as something upon which an operation or act is performed. Brittas Guesthouse and the initial tree house built for the Treelover documentary were essential material resources for the co-creation of value and meanings related to Treehotel. In the case of Icehotel, the actions, practices, and performances of employees, artists, and customers on ice and snow as construction materials contribute to developing new skills and knowledge, which in turn support the co-creation of value and meanings for the firm and its stakeholders.

Our analysis also draws attention to how these firms and their stakeholders act upon this value and meanings by incorporating them in stories and narratives. Indeed, as some scholars have suggested, storytelling can be a powerful tool for co-creation (e.g. Gebhardt, Carpenter, & Sherry Jr., 2006; Peñaloza & Mish, 2011). Both firms use their websites and social media channels to communicate their business stories, principles, and practices. In addition, storytelling is used interactively with other stakeholders, opening avenues for further co-creation of value through shared meanings in different sociocultural contexts (Helkkula et al., 2018). For example, journalists, travel bloggers, and even scholars have played an essential role in the co-creation of value associated with an environmentally responsible and sustainable world view. From this perspective, we can argue that these meanings are used to nurture not only local brand identities (see Schilar & Keskitalo, 2018) but also consumer cultures in the global marketplace (see Schau, Muñiz Jr., & Arnould, 2009).

Unexpected events

Our findings point to unexpected events in the form of challenges and opportunities that had an impact on co-creation. In the Icehotel, we identify two critical events, which changed the course of the firm. The first event was the rafting accident, which put on hold the main source of revenue of the firm. In the search for new business ideas, the firm's CEO ended up in the ice sculpture festival in Japan, where he built relations with local Japanese artists. This was the start of a co-creation process based on trial-and-error learning, in which the firm employees and Japanese artists in interaction with the Torne River develop the knowledge and skills needed to build ice and snow structures (see Miner et al., 2001).

The second event happened years later when ice and snow buildings and structures were part of the firm's offering. Confronted with a customer request for accommodation when all available cabins were booked, one of the managers, in collaboration with the customer, came up with the solution of using one of the ice buildings as an accommodation facility. For this solution to work, the customer, the firm, and the employees had to develop new knowledge, skills, and features for the ice and snow building that had not previously been considered. These include safety issues, sleeping practices, and the introduction of new materials such as sleeping bags and reindeer hides. In these improvisational activities, the material resources available in the firm and its operating environment became essential to the co-creation process (Cunha et al., 1999).

Improvisation is not only about challenges but also about unforeseen opportunities. This is the case of Treehotel, which saw an opportunity in the tree house left in the forest after the making of the documentary 'The Treelover'. When taking an improvisation approach to value co-creation, it unfolds as a series of dynamic interplays among different stakeholders, materials, spaces, and temporalities rather than an operationally managed process (Haanpää et al., 2016; Lugosi, 2014). Nevertheless, effective improvisation in value co-creation demands skills that allow firms to interpret the present through past experiences, and thus to enhance future performance (see Cunha et al., 1999; Haanpää et al., 2013; Haanpää et al., 2016). The owners of Treehotel seem to have the skills to read their place (past and present) and identify possibilities for developing new ideas and service concepts in close collaboration with their stakeholders (Brown & Eisenhardt, 1998). Indeed, they have the skills to look for stakeholders who can join them as partners in a project driven by commonly shared values and meanings (Peñaloza & Mish, 2011).

Implications and suggestions for future research

Although this study focuses on the data gathered from two northern Swedish tourism companies, it is supported by ethnographic research conducted over a period of several years. The two firm cases examined in this paper help us illustrate patterns of co-creation that we identified in a large number of small tourism firms operating in northern Finland and Sweden. In doing so, this study contributes to the discussion on co-creation in tourism by unpacking the dynamics of this theoretical concept. By drawing upon cultural marketing and organisational improvisation, we have identified three aspects of co-creation, which have received little attention in tourism management: stakeholders, material resources, and unexpected events.

Regarding stakeholders, this work is consistent with the notion of market co-creation introduced by cultural marketing scholars Peñaloza and Mish (2011). Our study challenges the view prevailing in tourism management studies that co-creation is just a strategic tool for acquiring and generating knowledge through customer participation. Indeed, as we show in this paper, co-creation in small tourism firms is embedded in a dynamic web of stakeholder relations and interactions. Not only customers, but also employees, local community members, governmental organisations, the media, and other stakeholders are legitimate actors in the co-creation of value. Our findings also acknowledge non-human actors as key stakeholders in co-creation. The forest and river, as non-human actors in this study, play a role in co-creation processes and the development of knowledge and skills. Considering nature as a stakeholder, we contribute to paving the way for studying human and non-human relations in co-creation in the field of marketing and tourism studies in particular (Bertella, 2014; Lugosi, 2018; Peñaloza & Mish, 2011).

In relation to material resources, we already see a shift in the status of the natural environment from being a material resource to becoming a stakeholder. In line with Schilar and Keskitalo (2018), we found that the emotional ties of the owners/managers of Treehotel and Icehotel played a key role in how these firms acted on the material resources available to them. Our findings also show how meanings attached to Treehotel and Icehotel are turned into material resources employed by the media and tourism organisations to create meanings for brands and consumer communities around the world. This has become possible due to the role of social media, in particular, and also due to the deterritorialisation of cultural contexts of value co-creation (Helkkula et al., 2018).

Concerning unexpected events, the study draws attention to the improvisational nature of co-creation in tourism. External factors in the form of challenges and opportunities can abruptly influence the dynamics of co-creation. In the cases included in this study, we recognise the role of the place-related events and the introduction of new material resources in shaping the co-creation of Treehotel and Icehotel (see Schilar & Keskitalo, 2018). An improvisation approach contributes to unveiling the role of time, space, emotions, feelings, and learning in value co-creation (Haanpää, 2017; Haanpää et al., 2016). Furthermore, it helps us understand and appreciate different forms of knowledge and knowing. Indeed, the unexpected events taking place in Treehotel and Icehotel demonstrate that the knowledge and skills of customers are equally important as the knowledge and skills of other stakeholders (García-Rosell et al., 2007; Peñaloza & Mish, 2011). In this sense, stakeholders other than customers are not only legitimate but also able to contribute with relevant knowledge and skills to co-creation processes.

Based on our findings, we argue that co-creation is an ongoing process embedded in the everyday practices of tourism firms and influenced by stakeholder interactions. By drawing attention to the expression 'dig where you stand', we want to stress the contextual nature of value co-creation in tourism (see Haanpää et al., 2016; Helkkula et al., 2018). By relying on two firm stories, we explore value co-creation through different contexts (e.g. time, space, material, and practices). In particular, our study shows how these two firms creatively rely upon their stakeholders, places, and material resources available to them.

We acknowledge that our study has limitations that point to the need for further research. One limitation is the focus on only two firms and one particular geographical location. It is not possible to draw conclusions that similar co-creation processes may arise in other tourism firms and regions. As we pointed out, co-creation is contextual in nature, influenced by material resources and stakeholder relations that exist within a given time, space, and sociocultural context. Although the stories of Treehotel and Icehotel offered insights into how value is co-created in the small tourism firm context, it raised questions that call for future studies.

Considering that tourism can also be viewed as a destructive force (Schilar & Keskitalo, 2018), future research should also explore value co-destruction in a tourism context. Following Plé and Chumpitaz Cáceres's work (2010), studies could focus on examining cases of how value can also be co-destroyed through stakeholder interactions and the accidental or intentional misuse of stakeholder relations and material resources. In line with Vera and Crossan (2004), it is also important to differentiate between how improvisation occurs and what it takes to do it well. In this study, we have illustrated the how by focusing on two tourism companies, and future research could look at cases where improvisation has not led to positive results. This will help us to better understand the skills and capabilities that play a role in turning unexpected events into a trigger for service development and innovation.

Disclosure statement

No potential conflict of interest was reported by the authors.

Funding

This work was supported by the European Social Fund under Grant S12344.

ORCID

José-Carlos García-Rosell http://orcid.org/0000-0001-5649-8838
Minni Haanpää http://orcid.org/0000-0003-3621-6957

References

Ateljevic, I., & Doorne, S. (2000). Staying within the fence: Lifestyle entrepreneurship in tourism. *Journal of Sustainable Tourism, 8*(5), 378–392.

Atkinson, R. (2002). The life story interview. In J. F. Gubrium & J. A. Holstein (Eds.), *Handbook of interview research: Context & method* (pp. 121–140). London: Sage.

Barrett, F. (1998). Creativity and improvisation in jazz and organizations: Implications for organizational learning. *Organization Science, 9*(5), 543–555.

Bertella, G. (2014). The co-creation of animal-based tourism experience. *Tourism Recreation Research, 39*, 115–125.

Bredvold, R., & Skålen, P. (2016). Lifestyle entrepreneurs and their identity construction: A study of the tourism industry. *Tourism Management, 56*, 96–105.

Brown, S. L., & Eisenhardt, K. M. (1998). *Competing on the edge: Strategy as structured chaos*. Boston, MA: Harvard Business School Press.

Cabiddu, F., Lui, T.-W., & Piccoli, G. (2013). Managing value co-creation in the tourism industry. *Annals of Tourism Research, 42*, 86–107.

Campos, A. C., Mendes, J., do Valle, P. O., & Scott, S. (2018). Co-creation of tourist experiences: A literature review. *Current Issues in Tourism, 21*(4), 369–400.

Cova, B., Dalli, D., & Zwick, D. (2011). Critical perspectives on consumers' role as 'producers': Broadening the debate on value co-creation in marketing processes. *Marketing Theory, 11*(3), 231–241.

Crossan, M., Cunha, M. P., Vera, D., & Cunha, J. (2005). Time and organizational improvisation. *Academy of Management Review, 30*(1), 129–145.

Crossan, M., & Hurst, D. (2006). Strategic renewal as improvisation: Reconciling the tension between exploration and exploitation. *Ecology and Strategy: Advances in Strategic Management, 23*, 273–298.

Cunha, M. P., da Cunha, J. V., & Kamoche, K. (1999). Organizational improvisation: What, when, how and why. *International Journal of Management Reviews, 1*(3), 299–341.

Cunha, M. P., Neves, P., Clegg, S. R., & Rego, A. (2015). Tales of the unexpected: Discussing improvisational learning. *Management Learning, 46*(5), 1–19.

Dredge, D., & Jenkins, J. (2011). New spaces of tourism policy and planning. In D. Dredge & J. Jenkins (Eds.), *Stories of practice: Tourism policy and planning* (pp. 1–12). Farnham, UK: Ashgate.

Echeverri, P., & Skålén, P. (2011). Co-creation and co-destruction: A practice-theory based study of interactive value formation. *Marketing Theory, 11*(3), 351–373.

Edvardsson, B., Tronvoll, B., & Gruber, T. (2011). Expanding understanding of service exchange and value co-creation: A social construction approach. *Journal of the Academy of Marketing Science, 39*(2), 327–339.

Eisenhardt, K. M. (1997). Strategic decision making as improvisation. In V. Papadakis & P. Barwise (Eds.), *Strategic decisions* (pp. 251–257). Clevedon: Chanel View Publications.

Eriksson, P., & Kovalainen, A. (2008). *Qualitative methods in business research*. London: Sage.

García-Rosell, J. C., & Haanpää, M. (2017). Developing practice and theory together: Reflecting on a tourism development and research project in Finnish Lapland. *Loisir et Société (Society and Leisure), 40*(2), 284–301.

García-Rosell, J. C., Haanpää, M., Kylänen, M., & Markuksela, V. (2007). From firms to extended markets: A cultural approach to tourism product development. *Tourism Review, 55*(4), 445–459.

García-Rosell, J. C., Kylänen, M., Pitkänen, K., Tekoniemi-Selkälä, T., & Vanhala, A. (2015). *Tourism co-creation handbook*. Retrieved from http://matkailu.luc.fi/Product-Development/Homepage.

Gebhardt, G. F., Carpenter, G. S., & Sherry Jr., J. F. (2006). Creating a market orientation: A longitudinal, multifirm, grounded analysis of cultural transformation. *Journal of Marketing, 70*(4), 37–55.

Grönroos, C. (2012). Conceptualizing value co-creation: A journey to the 1970s and back to the future. *Journal of Marketing Management, 28*(13–14), 1520–1534.

Grönroos, C., & Voima, P. (2013). Critical service logic: Making sense of value creation and co-creation. *Journal of the Academy of Marketing Science, 41*(2), 133–150.

Haanpää, M. (2017). *Event co-creation as choreography: Autoethnographic study on event volunteer knowing*. Rovaniemi: Lapland University Press.

Haanpää, M., García-Rosell, J. C., & Kyyrä, S. (2013). Ennakoiva tuotekehitys matkailussa [Foresight approach to product development in tourism]. In S. Veijola (Ed.), *Matkailututkimuksen lukukirja [A guide to tourism studies]* (pp. 102–114). Rovaniemi: Lapland University Press.

Haanpää, M., García-Rosell, J. C., & Tuulentie, S. (2016). Co-creating places through events: The case of a tourism community event in Finnish Lapland. In A. Jepson & A. Clarke (Eds.), *Managing and developing communities, festivals and events* (pp. 34–49). New York: Palgrave MacMillan.

Hammersley, M., & Atkinson, P. (1996). *Ethnography: Principles in practice*. London: Routledge.

Helkkula, A., Dube, A., & Arnould, E. (2018). The contextual nature of value and value co-creation. In S. L. Vargo & R. F. Lush (Eds.), *The Sage handbook of service-dominant logic* (pp. 118–132). Thousand Oaks, CA: Sage Publications.

Hjalager, A.-M., & Nordin, S. (2011). User-driven innovation in tourism—A review of methodologies. *Journal of Quality Assurance in Hospitality & Tourism*, *12*(4), 289–315.

Holbrook, M. B. (1999). *Consumer value: A framework for analysis and research*. London: Routledge.

Holbrook, M. B. (2006). ROSEPEKICECIVECI versus CCV: The resource-operant, skills-exchanging, performance-experiencing, knowledge-informed, competence-enacting, co-producer-involved, value emerging, customer-interactive view of marketing versus the concept of customer value: "I Can Get It for You Wholesale". In R. F. Lusch & S. L. Vargo (Eds.), *The service-dominant logic of marketing: Dialog, debate and directions* (pp. 208–223). Armonk, NY: M. E. Sharpe.

Jorgensen, D. L. (1989). The methodology of participant observation. In D. L. Jorgensen (Ed.), *Participant observation* (pp. 12–26). Thousand Oaks, CA: Sage.

Kamoche, K., & Cunha, M. P. (2001). Minimal structures: From jazz improvisation to product innovation. *Organization Studies*, *22*(5), 733–764.

Kyriakopoulos, K. (2011). Improvisation in product innovation: The contingent role of market information sources and memory types. *Organization Studies*, *32*(8), 1051–1078.

Lugosi, P. (2014). Mobilising identity and culture in experience co-creation and venue operation. *Tourism Management*, *40*, 165–179.

Lugosi, P. (2018). Disruptive ethnography and knowledge co-creation. In C. Ren, G. Jóhannesson, & R. van der Duim (Eds.), *Co-creating tourism research* (pp. 55–72). London: Routledge.

Miner, A. S., Bassoff, P., & Moorman, C. (2001). Organizational improvisation and learning: A field study. *Administrative Science Quarterly*, *46*, 304–337.

Moisander, J., & Valtonen, A. (2006). *Qualitative marketing research: A cultural approach*. London: Sage.

Payne, A., Storbacka, K., & Frow, P. (2008). Managing the co-creation of value. *Journal of the Academy of Marketing Science*, *36*(1), 83–96.

Peñaloza, L., & Mish, J. (2011). The nature and processes of market co-creation in triple bottom line firms: Leveraging insights from consumer culture theory and service dominant logic. *Marketing Theory*, *11*(99), 9–34.

Peñaloza, L., & Venkatesh, A. (2006). Further evolving the new dominant logic of marketing: From services to the social construction of markets. *Marketing Theory*, *6*(3), 299–316.

Plé, L., & Chumpitaz Cáceres, R. (2010). Not always co-creation: Introducing interactional co-destruction of value in service-dominant logic. *Journal of Services Marketing*, *24*(6), 430–437.

Pongsakornrungsilp, S., & Schroeder, J. E. (2011). Understanding value co-creation in a co-consuming brand community. *Marketing Theory*, *11*(3), 303–324.

Prahalad, C., & Ramaswamy, V. (2000). Co-opting customer competence. *Harvard Business Review*, *78*, 79–90.

Prahalad, C., & Ramaswamy, V. (2004). Co-creation experiences: The next practice in value creation. *Journal of Interactive Marketing*, *18*(3), 5–14.

Prebensen, N. K., Chen, J. S., & Uysal, M. (2014). *Creating experience value in tourism*. Oxfordshire: CABI.

Rantala, O. (2011). An ethnographic approach to nature-based tourism. *Scandinavian Journal of Hospitality and Tourism*, *11* (2), 150–165.

Rantala, O., & Mäkinen, M. (2018). Engaging with wind shelters. In C. Ren, G. Jóhannesson, & R. van der Duim (Eds.), *Co-creating tourism research* (pp. 131–146). London: Routledge.

Rihova, I., Buhalis, D., Moital, M., & Gouthro, M. B. (2013). Social layers of customer-to-customer value co-creation. *Journal of Service Management*, *24*(5), 553–566.

Roeffen, D., & Scholl-Grissemann, U. (2016). The importance of customer co-creation of value for the tourism and hospitality industry. In R. Egger, I. Gula, & D. Walcher (Eds.), *Open tourism. Tourism on the verge* (pp. 35–46). Berlin, Heidelberg: Springer.

Schau, H. J., Muñiz Jr., A. M., & Arnould, E. J. (2009). How brand community practices create value. *Journal of Marketing*, *73* (5), 30–51.

Schilar, H., & Keskitalo, C. H. (2018). Tourism activity as an expression of place attachment-place perception among tourism actors in the Jukkasjärvi area of northern Sweden. *Scandinavian Journal of Hospitality and Tourism*, *18*(S1), 42–59.

Storbacka, K., Frow, P., Nenonen, S., & Payne, A. (2012). Designing business models for value co-creation. In S. L. Vargo & R. F. Lusch (Eds.), *Special issue – Toward a better understanding of the role of value in markets and marketing. Review of marketing research* (Vol. 9, pp. 51–78). Bingley: Emerald Group Publishing.

Valtonen, A. (2009). Small tourism firms as agents of critical knowledge. *Tourist Studies*, *9*(2), 127–143.

Vargo, S., & Lusch, R. (2004). Evolving to a new dominant logic for marketing. *Journal of Marketing*, *68*, 1–17.

Vargo, S., & Lusch, R. (2008). Service-dominant logic: Continuing the evolution. *Academy of Marketing Sciences*, *36*(1), 1–10.

Vera, D., & Crossan, M. (2004). Theatrical improvisation: Lessons for organizations. *Organization Studies*, *25*(5), 727–749.

Yachin, J. M. (2019). The entrepreneur-opportunity nexus: Discovering the forces that promote product innovations in rural micro-tourism firms. *Scandinavian Journal of Hospitality and Tourism*, *19*(1), 47–65.

The co-creation of diverse values and paradigms in small values-based tourism firms

Lucia Tomassini

ABSTRACT
Although small firms characterise the tourism sector, there is currently insufficient critical understanding of firms that do not define themselves in commercial terms but through the non-profit values they pursue and the ethical vision they are committed to. This paper refers to such firms with the expression 'values-based' and examines their co-creation of diverse values and paradigms. Through a social constructionist perspective, the research designs a qualitative narrative approach based on a number of lightly guided interviews with the owner-managers of small Italian tour operators: all members of the Italian Association of Responsible Tourism, committed to an ethical vision of tourism, and operating in developing countries. The study implements a narrative analysis that reveals common patterns in the way participants make sense of their driving values and co-create – together with their collaborators, partners and customers – an ethical vision of tourism and driving-values in respect to global challenges. The paper proposes a critical stance on how small values-based tourism firms understand and narrate the process of co-creating diverse values and paradigms. This study challenges several assumptions about how non-commercially oriented firms contribute to co-creating alternative values and paradigms in tourism with regards to development, growth, citizenship and entrepreneurship.

Introduction

This paper explores how small tourism firms that define themselves through the non-profit values they pursue and the ethical vision they are committed to, narrate the co-creation of diverse values and paradigms through their relationships with collaborators, partners and customers. Hitherto, small tourism firms have been understood mainly on the basis of a rudimentary polarisation – commercially oriented and non-commercially oriented – where non-commercially oriented firms have been identified as mostly lifestyle-oriented, i.e. as rejecting growth in order to pursue personal lifestyle choices (Ateljevic & Doorne, 2000; Thomas, Shaw, & Page, 2011). Since the term lifestyle-oriented is elusive and misses many of the nuances related to diverse driving values and goals, this study adopts the term 'values-based' to refer to small non-commercially oriented tourism firms that are committed to ethical goals and engaged in alternative proactive approaches to development, growth, and welfare.

These firms critically reflect part of the contemporary debate on global challenges and the development agenda, as well as the transition from consumerism to citizenship, where civil society and companies are called to commit themselves to the development agenda, take responsibility for others, and expand their moral circle (Easterly, 2006; Elkington, 1998; Klein, 2000; Singer, 2010, 2015). This study scrutinises how, within the tourism sector, these firms co-create and embody alternative ideological paradigms and approaches to development that build on the post-development theorisations of authors such as Illich (1971), Latouche (2004), Schumacher (1973) and Bauman (2000).

According to the literature, co-creation indicates – respectively, in the private and public sectors – the active participation of end-users as sources of product and service innovation (Von Hippel, 1989; Voorberg, Bekkers, & Tummers, 2015), and of citizens as part of the social innovation process (Voorberg et al., 2015). This study proposes a critical reflection on how small values-based tourism firms recount the co-creation of alternative values and paradigms through a dialectical, open, direct relationship with their collaborators, partners, and customers. On the one side, the study resonates with studies exploring how a networked community of stakeholders engaged in co-creation processes makes firms more innovative and competitive (Prahalad & Ramaswamy, 2000; Ramaswamy & Gouillart,

2010). On the other side, the study frames co-creation as expression of proactive citizenship, since firms and stakeholders appear to be subjects deeply engaged in making a difference in the debate on global issues, being moved by ideological critiques of the inequalities within the global economy.

The research uses a narrative approach whereby the researcher operates as a bricoleur conveying a number of crucial topics into a complex dataset of meaningful narratives to be analysed and interpreted. Through purposive sampling, the researcher identifies seven Italian tour operators operating in developing countries, and leads lightly-guided interviews with these firms' founders/owners, collecting a set of first-person accounts to be analysed via a multilevel narrative analysis. This approach offers the opportunity to connect different topics, correlate them with each other, and interpret the attached meanings. The way in which stories and narratives are linguistically and structurally constructed offers an interesting perspective to deepen the understanding of co-creation in dialectical and relational terms.

Alternative paradigms of development and growth, and the co-creation of diverse values in tourism

The tourism industry is permeated by values. While the majority of tourism appears inseparable from contemporary consumerism and neo-liberal capitalism (Hall, 2011), this paper explores a set of diverse values rooted in alternative paradigms of development, growth and consumption.

The theoretical framework examined in this section, therefore, aims to shed light on how small values-based tourism firms construct their identity and make sense of their driving values. The theoretical framework provides a novel critical stance within tourism studies since it paints a sense of belonging and frankness between firms and stakeholders that implies the co-creation of a common ground of values while constructing firms' identity around taking responsibility for others. The section moves from authors such as Ivan Illich and Serge Latouche and their theories of development, de-growth and conviviality (Illich, 1971, 1973; Latouche, 1993, 2004), Leopold Kohr's and Friedrich Schumacher's theories of 'small is beautiful' (Illich, 1997; Kohr, 1986; Schumacher, 1973), the role of civil society in the shift from 'consumerism to citizenship' (Elkington, 1998; Klein, 2000), to forms of pragmatic altruism (MacAskill, 2015; Singer, 2010, 2015) and *parrhesiastic* 'truth-telling' as assumptions of responsibility (Foucault, 1983; Lardinois, 2011; Sluiter & Rosen, 2004).

Thinking, then, of the relationship between development and growth, from the second half of the twentieth century, 'development' has been a value-laden word with a complex history in combination with the goal of poverty reduction. Already in the 1970s, however, Illich stated the necessity to re-think a one-size fits all approach to development (Illich, 1971). Illich raised doubts about the effective relationship between institutionalised services – such as schools and hospitals – and the underpinning values of education and health. Moreover, Illich highlighted the potential of a society based on conviviality and the consciousness of its limits in terms of growth, and provided a new perspective through which to look at development, introducing the concept of de-growth as an alternative economic model based on the awareness of those limits (Illich, 1973). Illich stressed the importance of acquiring 'Tools for conviviality' to make people free from a technocratic society where citizens are conceived of mainly as consumers, and where they are deprived of full understanding, knowledge and human relationships in the production/consumption process (Illich, 1973).

Later, Latouche investigated the idea of de-growth by saying:

> it might be more accurate and less alarming if we replaced the word de-growth with 'non-growth'. We could then start talking about 'a-growthism', as in 'a-theism'. After all, rejecting the current economic orthodoxy means abandoning a faith system, a religion. To achieve this, we need rigorously to deconstruct the matter of development. The term *development* has been redefined and qualified so much that it has become meaningless. (Latouche, 2004, vol. 11)

In this context, development research has shown increasing interest in Leopold Kohr's research subverting conventional economic wisdom by criticising the cult of bigness and promoting the concept of a human scale of small community life (Illich, 1997). Schumacher, meanwhile, proposed the idea of 'smallness within bigness' as a sort of decentralisation where a large organisation has to work like a related group of small organisations (Schumacher & McKibben, 2010). These perspectives have echoes in the latest research investigating the failures in the development agenda. Easterly criticises big long-term aid plans and supports the idea of an aid system based on smaller, short-term, bottom-up, and randomised interventions (Easterly, 2006).

During the last two decades, these debates about development have energised a significant role for civil society in terms of a proactive concept of citizenship where both individual citizens and organisations are called to take responsibility for others and for the environment (Elkington, 1998; Klein, 2000, 2017). This could be conceptualised as 'consumerism versus citizenship' (Klein, 2000, p. 439) and as the shift from 'global consumer to world citizen' (Elkington, 1998, p. 151). In order to outline the role of civil society and bottom-up processes in taking responsibility for others and the environment, it is worth noting the birth of movements like the Effective Altruism movement, which promotes the enlargement of one's private moral circle in favour of a personal pragmatic commitment towards others (MacAskill, 2015; Singer, 2010, 2015).

These new forms of proactive citizenship, personal commitment and assumption of responsibilities seem to resonate also with the classical Greek notion of *parrhesia*, which literally means 'all-telling' and – by extension – 'to speak freely' (Foucault, 1983; Lardinois, 2011; Sluiter & Rosen, 2004). According to this notion – discussed by Michael Foucault in a set of lectures at Berkley University (Foucault, 1983) – the *parrhesiast* is a citizen who is usually inferior to his/her recipient from the point of view of economic and political power. He/she is a 'truth-teller' who speaks openly and freely about inequalities and lack of democracy in order to convey the 'truth' and a deeper understanding to others.

This critical debate offers a lens through which co-creation of diverse alternative values and paradigms in the tourism industry can be understood; in that these tourism-related values and paradigms can be seen as both extending and reacting to this debate about the nature of development and the role of the citizen. The next section presents the narrative approach adopted by this study to examine the participants' recounting of their co-creation of alternative values and paradigms in terms of individual sense-making and understanding.

Methodology

This study is framed into a social constructionist paradigm and implements a narrative approach to data collection and data analysis by exploring small values-based tourism firms through the voice of their owner-managers (Gee, 1991; Labov & Waletzky, 1997; Riessman, 1993). The paper uses the term 'owner-manager' due to the acknowledged multitasking role of small firms' owners (Dewhurst & Horobin, 1998; Sampaio, Thomas, & Font, 2011). While disclosing participants' sense-making and beliefs, this methodological approach contributes to the examination of the co-creation of values that takes place dialectically between firms' founders-owners and their network of stakeholders. This study reveals how co-creation is not only about the creation of 'things', but also about co-created meanings through social relationships, interactions and dialectical approaches (Ind & Coates, 2013; Von Hippel, 1989).

The researcher turned to the owner-managers of seven Italian tour operators, identified through a purposive sampling among the members of the Italian consortium AITR – the Italian Association of Responsible Tourism. The participants are middle age owner-managers of Italian tour operators operating mainly in developing countries. The researcher arranged an interview via Voice over Internet Protocol (VoiP) – Skype audio call – given that the participants were displaced across different parts of Italy. Each interview lasted between 45 and 60 minutes. The researcher involved participants – four men and two women – in lightly-guided, open-ended interviews, eliciting the conversation through a set of open and flexible questions and stimulating the relationship of narrator/listener instead of interviewed/interviewer (Chase, 2005; Johansson, 2004; Riessman, 1993).

This narrative inquiry resulted in a set of first-person accounts in Italian – the researcher's mother tongue – which needed to be carefully transcribed, analysed, and then finally translated into English. Narratives are representations, therefore different analytical approaches can be combined to interpret them (Carless & Sparkes, 2008; Riessman, 1993; Smith & Sparkes, 2008). In this research, a structural approach scrutinised the interviewee's personal narratives in storified form. The open-ended interview kept – as an entry point – the question 'Would you tell me in your own words the story of your firm?' This produced a personal narrative in a storified form – having a plot with a beginning, middle and an end – that largely matched Labov's criteria for structural analysis, according to which a narrative is examined in terms of the chronological and causal relationships between its different units (Labov, 2010; Labov & Waletzky, 1997; Riessman, 1993). The researcher organised the text in numbered lines, distinguishing them according to the central idea they conveyed. Adopting the names used by William Labov, the text is organised by dividing it according to structural categories, namely: abstract, orientation, complication, evaluation, resolution, and coda, each one with a functional property (Labov, 2010; Labov & Waletzky, 1997).

The open-ended lightly guided interviews also generated stretches of talk largely recounting opinions, feelings and points of view not matching with the criteria of a structural functional approach. The study identified and analysed these through a different approach derived from the linguistic approach to the narrative of James Gee (Gee, 1991; Riessman, 1993). According to Gee, units of speech can be identified, paying attention to how the narrative is told, and focusing on linguistic and extra-linguistic features such as pauses, elongated vowels, emphasis, pitch, repetitions, discourse markers, pitch glides, and non-lexical elements. These provide the key to individuate and group different lines of text and stretches referring to the same topic. These units, referred to by Gee as Stanzas or Strophes (Gee, 1991), are thematically coherent and tightly sequenced and allow the analysis of narratives not in a storied form (Riessman, 1993). The researcher therefore organised these non-storified texts in numbered 'lines', 'stanzas' 'strophes' and 'parts', differentiating them according to the theme – 'image' – they focus on. The adoption, adaptation and combination of these two approaches allowed the analysis and interpretation of narrative data, conveying into research findings.

Results

Besides conveying a personal point of view, participants' narratives reveal a form of co-creation in the way that they make sense of their driving values, beliefs and visions. Co-creation is here examined in relational terms since participants narrate and make sense of alternative values and paradigms, as outcomes of an open, dialectical relationship with their collaborators, partners and customers (Ind & Coates, 2013; Von Hippel, 1989). The research findings depict a realm of owner-managers perceiving themselves as part of a community of people – collaborators, partners and customers – sharing the same vision and having similar goals. Although each participant is an individual case with his/her own attitude, characteristics and understanding; the narrative analysis discloses common threads in how participants co-construct diverse values like smallness (Schumacher & McKibben, 2010), de-growth (Latouche, 2004), artisanal production (Rickly-Boyd, 2012), disruptive entrepreneurship (Schumpeter, 2000), empathy (Singer, 2010, 2015), and a parrhesiastic approach within the tourism industry (Foucault, 1983).

This section presents a set of key driving-values recounted by the firms' owner-managers: from smallness, de-growth and the artisanal dimension of production, to empathy and parrhesia, and a diverse sense of disruptive entrepreneurship. The way in which participants make sense of their firms, their network of stakeholders and their driving values, builds on the work of authors exploring co-creation as a participative process where people and organisations generate and develop meanings together (Ind & Coates, 2013; Ryan, 2012; Voorberg et al., 2015). The research findings present a culture of co-creation that is not only wider and more diverse than the managerial interpretation seems to suggest in terms of co-production of value and addition of extra value (Ind & Coates, 2013; Prahalad & Ramaswamy, 2000) but that also resonates with a sense of greater and deeper citizen participation (Ryan, 2012). The participants' narratives, therefore, reveal how meanings largely emerge through social and communicative processes since meaning itself is dialogic and is always co-created (Ind & Coates, 2013). Hence, it arises through a process of value co-creation that is largely inclusive, relational and collaborative. In summary, small values-based tourism firms are recounted as the result of a negotiation of meanings within a networked social and economic world (Magala, 2009; Merz, He, & Vargo, 2009).

Smallness, de-growth and the artisanal dimension of production

Participants' narratives convey the negotiation of meanings that took place within a wider circuit of subjects: collaborators, partners, customers, and the umbrella association they belong to as members. The co-creation process mainly emerges from the recounting of their personal and professional history, and from the story of the firm's origin. Participants seem to understand their identity construction process as a long-term co-creative process built on a cultural mediation and relationship not only with stakeholders, but also with an alterity represented by external subjects. For instance, participant 1 narrates the relationship with groups belonging to wider networks in order to absorb, discuss and co-create ideas:

> *Participant 1, Firm 1*
>
> *(Part 1, Strophes 1–5, Stanzas 3–16)*
>
> 10 the idea was to create a bridge between these different cultures
>
> 11 and the travel was one of the ways to escape the commercial dimension through realities that had a meaning, and so we begin
>
> 17 but for us dealing with responsible tourism means quite another idea: first of all it means criticising tourism, it means then discussing tourism
>
> 61 an idea shared with another small group of people
>
> 63 because we adhere to the Tourism European Network, and that is decisive
>
> 64 because they offer us a whole range of crucial points, of ideas; but they really water us
>
> 65 with what was the related culture that has been known in Europe and around the world for fifteen years

Participant 2 stresses the importance of the relationship and cultural collaboration with ethnic minorities to co-create not only a meaningful identity, but also a novel product with an added value:

Participant 2, Firm 2

(Part 1, Strophe 1, Stanza 1)

02 because we stem out from the idea of having some travel in collaboration with ethnic minorities, so a somewhat special thing of its kind

03 and then, from there, we soon realised that it was truly an asset having these links directly on site, with trustworthy people, with whom we worked for other projects

Throughout the narratives, participants' negotiation of meanings develops into an explanation of alternative driving-values and paradigms like smallness, conviviality and de-growth. Participants' interest in smallness appears to be the result of an identity dialectically co-constructed through the time, and juxtaposed with the failures and weaknesses of mainstream tourism and the contemporary economic model (Butcher, 2008; Singh, 2012). The interpretation of such findings is helped by authors investigating how small-scale operations and tourism products are largely presented as a means to enhance, not only local development, but also a feeling of belonging, ownership and humanistic principles among the diverse stakeholders (Butler, 2011; Harrison, 2011; Schumacher, 1973; Weaver, 2011). For instance, Participant 4 frames the interest in smallness both as a firm characteristic deeply intertwined with ethical and moral choices and as the outcome of the relationship with the firm's customers, who are in turn interested in smallness:

Participant 4, Firm 4

(Part 3, Strophe 7, Stanzas 25–27)

91 we reject, for instance, in the hospitality part, all the great hotel companies both because of, well, moral reasons, let's put it this way, but also because of commercial characteristics

92 I mean ours are groups are small,

93 people, what they appreciate more, is maybe this almost familiar atmosphere that is present during the holiday

99 but very often, well, it happens, where possible, to organise an experience with host families or, however, almost always, we choose small hotels, small facilities, that's it

100 Because it's the dimension that fits better our kind of travels, and traveller

Smallness and de-growth – this latter understood as 'non-growth', 'a-growth', as a lack of interest in growth both in the firm and in the work team – are narrated as unavoidable characteristics of an artisanal, tailor-made production that creates uniqueness and value-laden imperfections. According to these narratives, pursuing smallness and de-growth comes from a two-way connection between the firm and its stakeholders – collaborators, employees, local partners and customers –; in order to listen to their needs and inputs and take care of them. Participant 1 presents de-growth as a search for quality:

Participant 1, Firm 1

(Part 4, Strophe 12, Stanza 38)

180 then it is actually my colleague and I in the office

181 because among other things we chose not to grow because

182 I'll tell you why later and this is a very specific thing

183 but, it means quality to us

And part of being travel artisans with a model of production and consumption that values authenticity, time, and carefully listening to people:

Participant 1, Firm 1

(Part 3, Strophe 11, Stanzas 34–35)

159 so we can do this because we are travel artisans,

160 and it comes from direct experience,

161 from direct contacts with people,

162 that today are often also via Skype, or email properly,

163 but they are also straight direct.

164 this is something that I claim as a very specific peculiarity of Firm 1

(Part 4, Strophe 14, Stanza 44)

212 we care that the guide is able to translate encounters happening by chance, to visit reality and to talk with people,

213 and maybe have dinner

214 and, not to care for the time elapsing, but to be there right? Do you understand?

Participants retrace and welcome the aura surrounding artisanal work (Benjamin, Jennings, & Doherty, 2008; Rickly-Boyd, 2012). 'Aura' is a human experience, an engagement, and a desire for the uniqueness connected with the authenticity of the experience. 'Aura' is embedded in ritual and traditions, and it is also a matter of the time needed to produce and consume an experience (Benjamin et al., 2008; Rickly-Boyd, 2012). In tourism, this resonates with a predominantly relational authenticity of experience (Rickly-Boyd, 2012, p. 274). Participants understand themselves and their firms, as subjects striving to have the proper time to produce their products and to establish satisfying human relationships able to co-create meanings and added value, as Participant 7 highlights:

> *Participant 7, Firm 7*
>
> *(Part 4, Strophes 13–14, Stanzas 34–36)*
>
> 148 so it means: our work, which is very artisanal, really homemade,
>
> 149 even though, you make lots of mistakes, a lot of things, that could be done in a better way, for sure, and so on,
>
> 150 but, it's still an artisanal work,
>
> 151 well you know very well that, if you buy an Ikea piece of furniture, it might be beautiful, it's cheap
>
> 152 but, of course, the same piece of furniture made by an artisan, maybe has some flaws, but that's because it's a unique piece, right?
>
> 153 but it remains a unique piece,
>
> 154 and it's really different
>
> 155 and this, probably, well, can be judged both as a flaw, and as a virtue,
>
> 160 I wanted a strong, artisanal, impact to stay, a very personalised relationship also with our customers,
>
> 161 that I still struggle to call them like this,
>
> 162 I call them participants,
>
> 163 I like it more,

A narrative understanding of these findings discloses features belonging to a borderless, porous movement, rather than to organisations. Participants largely introduce themselves as global citizens and as a part of civil society. Firms are recounted as subjects valuing cultural exchanges together with meaningful and relational co-creation processes in opposition to the liquefaction of a modern society permeated by fragmentation, impermanence and discontinuity (Bauman, 1998, 2000; Franklin, 2003; Lee, 2005).

Empathy and parrhesia

Participants disclose how they largely understand co-creation as the capacity to establish a frank, open, empathic, and transparent relationship with their stakeholders (sometimes too much so) thanks to effective two-way communication. According to their narratives, the capacity to co-create values and paradigms emerges from a process of deep and frank listening and of taking care of others. This conveys to the company not only a sense of belonging to a bigger community of global citizens, but also innovation and extra value (Prahalad & Ramaswamy, 2000; Ramaswamy & Gouillart, 2010). For instance, Participant 2 presents this frank relationship as a challenging search for an effective balance:

> *Participant 2, Firm 2*
>
> *(Part 4, Strophe 8, Stanzas 21–22)*
>
> 138 but for us, effectiveness is the result we get with travellers, I mean, and then with the local community
>
> 140 so always having a balance between positive feedback by travellers and by local hosts; this because it's important for us to be effective as a tour operator, both from the customer's point of view and from the supplier's point of view of.
>
> 141 At the risk of being sometimes maybe too sincere but I prefer too much sincerity than too less

Participant 6 as the capacity to give transparent, direct answers:

> *Participant 6, Firm 6*
>
> *(Part 7, Strophe 17, Stanza 38)*
>
> 141 I think that effectiveness, is succeeding, anyway, in giving answers to your interlocutors, both customers, and, well, potential customers
>
> 142 giving it to them immediately
>
> 143 but giving answers that are directed to their question,
>
> 144 I mean, do not stammering, and surely be straightforward,

Then, throughout their narratives, participants' convert this direct and transparent communication into empathy and parrhesia, mainly interpreted as manifestations of an open dialectical co-creation process. Both empathy and parrhesia can be understood as driving-values since participants make sense of themselves firstly as global citizens and free-speaking individuals and, secondly, as the founders-owners/managers of their firms. Participant 5 frames this in terms of a commonality of ideals:

> *Participant 5, Firm 5*
>
> *(Part 2, Strophe 5, Stanzas 16–17 and 24)*
>
> 68 so essential is the choice of the partner
>
> 70 I mean, the stronger this commonalities of ideals is, the more legs the relationship has
>
> 72 so, then, what is this commonality?

73 in my opinion, it's related to the general trend of the market,

99 I mean, again, finding, sharing with a local partner, who then manages a good part of the trip, a goal that is not only the trip itself, but it is also a vision of how it is the development, for example, of their own country

100 I mean, it seems now somewhat ideological this way, then, it translates in something real

This interpretation of the research findings dialogues not only with authors exploring the role of civil society in the global economy and the shift from consumerism to citizenship (Elkington, 1998; Klein, 2000), but also with the literature reviewed for this study about the role of *parrhesiastes* in ancient Greek society and the implications of the 'truth-teller' role in contemporary society and politics (Colclough, 1999; Foucault, 1983; Lardinois, 2011; Novak, 2006; Sharpe, 2007; Sluiter & Rosen, 2004). Since *Parrhesiastes* appear not to have any doubts about their possession of 'the truth' and of the moral qualities enabling them to tell it openly, participants' open, frank attitude is mainly understood as part of their truth-telling role. An example is Participant 3's reasoning on responsible tourism:

Participant 3, Firm 3

(Part 5, Strophe 16, Stanzas 66–68)

246 For example I tell you something technical: that the biggest part of the products of responsible tourism, I mean travels, are not conceived for distribution

247 I mean, they do not have a sufficient margin to hold an intermediation that goes further than direct sales

248 they are just not made to be in an agency

249 this is wrong!

250 it is wrong because you rule out what is the real target market

251 because we don't have to sell to people sharing our opinions

252 we must sell responsible tourism to people who never experienced it

Moreover, there is a sense of conviviality that permeates the narration of relationships that are human-based rather than professionally oriented. This resonates with Illich's critical studies about moving beyond institutionalised relationships and rigid hierarchical roles (Illich, 1971, 1973). Participants' narratives disclose a blend of conviviality, empathy and altruism that Participant 1 presents as the capacity to build a networked responsive co-creative community of stakeholders:

Participant 1, Firm 1

(Part 6, Strophe 22, Stanza 74)

365 well it luckily happens that small group of people who are our guides, they understood this issue very well

366 and so they speak with me very often; they are the ones taking the initiative

367 they go around and find a project we knew nothing about

368 they get involved, they take the group there and, in the end, they're all very happy, aren't they?

369 and then that project remains in our contact list

370 this doesn't happen only to our, let's say, professional guides, it also happens to individuals who had their private trips organised

371 Because some of them become friends

372 and send me lots of feedback pages: "we went, we met: these are great to have an encounter as it should be".

Participant 2 as the ability to establish empathic connections leading to co-creation:

Participant 2, Firm 2

(Part 2, Strophe 3, Stanza 7)

32 we were born very sector-based, right? With this type of local contacts very close to us, we immediately set up as a tour operator for individual journeys

33 and this has also changed our growth path, with a feedback very close to the host community

36 and with a situation that's also very empathic in the customer care since customers become almost travel mates even if you are not travelling with them, right?

37 therefore you stay with them a lot more

38 and this became, let's say, somehow our feature

This enduring search for open communication, truth, accountability and transparency reveals subtle connections with the Effective Altruism movement, a philosophy and social movement seeking to provide the greatest possible positive impact in improving the global issues and development challenges facing the world (Singer, 2010, 2015). Effective Altruism is a movement rooted in a utilitarian theoretical framework made up of the prioritisation of causes based on cost-effectiveness and impartiality, and which combines a search for effectiveness, accountability and transparency with an altruistic personal attitude (MacAskill, 2015; Singer, 2010, 2015). Basically, it is a form of altruism concerned about its own effectiveness. Participants in this study seem to understand themselves and their firms not only as empathic truth-teller subjects committed to altruistic moral obligations, but also as subjects determined to guarantee transparency, accountability and effectiveness concerning their impacts on stakeholders.

A diverse disruptive sense of entrepreneurship

Throughout their first-person accounts, participants present themselves not as replicating an entrepreneurial lifestyle aimed at 'staying within the fence' but as being innovative and creative entrepreneurs able to produce disrupting changes and results (Ateljevic & Doorne, 2000; Schumpeter, 1947, 2000; Shaw & Williams, 2004). Participants seem to understand the co-creation of alternative entrepreneurial values and paradigms mainly in terms of a small-scale, liquid, innovative, and resilient entrepreneurship imbued with Easterly's theorisation of 'searcher' entrepreneurs (Easterly, 2006). Facing inequalities and economic development challenges on a global scale, 'searchers' are a model of entrepreneurs working in a small-scale and randomised way to find out what the demand is within the market system. Thus, co-creation emerges from the participants' capacity to establish a deep and inclusive understanding of gaps in the market and stakeholders' needs and insights. According to Easterly's theorisation, 'searchers' take responsibility for their products, know how things go from the bottom and have an approach based on feedback, accountability and promptness (Easterly, 2006). Participants largely frame co-creation as a disruptive and inclusive entrepreneurship valuing stakeholders' contributions and external world stimuli, since these convey novel paradigms and successful outcomes. Participant 7 links his capacity to soak himself in the external world to the successful growth of his firm:

Participant 7, Firm 7

(Part 14, Strophe 43, Stanzas 100–102)

461 I'd like to add that, however, I am not hiding in my shell waiting for things to go a certain way

462 I live a lot in the outside world

463 well, I read up very much,

464 I watch what happens

468 Anyway, note that, however, we are constantly growing economically

470 And, also from the point of view of investments, and human resources, it is a very important thing to consider

Participant 4 narrates his disruptive co-creative entrepreneurship in terms of being professionally responsive to customers' diverse needs:

Participant 4, Firm 4

(Part 6, Strophe 18, Stanzas 74–76)

277 in my opinion the decision that saved us

278 it was the one of diversifying, in the sense that, in our field, it was very important

282 some years ago, we decided to start, I mean, we stepped in, our customers' shoes, but if people travel with us, and want to have a journey in a certain way, why should we send them only to Africa and Latin America?

283 So we started including Asia as well, and we started to do things in Europe

284 and so we started to make thematic areas: we began with educational trips, that didn't exist at the beginning, journeys on foot, honeymoon

Professionalism and entrepreneurship are largely characterised by a feeling of rebelliousness, and a maverick approach concerning the inequalities of the global economy. The rebel and maverick self-introduction is reminiscent of Schumpeter's conception of entrepreneurs as 'wild spirits' (Schumpeter, 1947, 2000). Nevertheless, participants' narratives picture a personal search for alternative paths and underground dimensions that is intertwined with a desire for inclusiveness and relational co-creation. As exemplified by Participant 3's story, participants largely understand themselves as 'rebellious' citizens disappointed by the global economy and willing to take the risk to promote change through a more inclusive entrepreneurial paradigm:

Participant 3, Firm 3

(Participant's story, Resolution)

60 indeed it was the first trip, in which, while having the same trip I had the first time, I actually discovered an underground dimension that I had never known.

61 And so that thing made me instantly feel, a sensation,

62 namely: that what I had just experienced as discomfort had a solution, and that it was also extremely interesting,

63 that is, if I'd known to all the people who had written to me before this trip, if I'd known if I had known all these realities I'd have told them!

65 And as they were many I've always been a pretty resourceful person,

66 the first thing I thought was: "now I make a website. Now"

67 and, instead of sitting there writing blah blah blah, every time, long and longer mail, I put the website reference

Participants present themselves as inexpert outsiders who – thanks to their systematic professional work and a strong community of stakeholders – have been able to establish a fruitful firm, as narrated by Participant 4:

Participant 4, Firm 4

(Part 4, Strophe 13, Stanzas 45–46)

164 because, anyway, when we started off in ninety-eight, I use the plural because we are a cooperative and I think what I tell, would be told by the others as well, not with my experience,

167 Now, we are, however, a reality with four working associates, basically, full time, and a whole series of occasional collaborators, in different countries

168 but, in short, we are a very small company, but we are a company, however, we are a cooperative, with its own budget:

169 last year we closed the budget with one million and seven hundred thousand euro incoming,

170 well we have been growing, especially, over the years

Participants' entrepreneurship deals with the search for development, tangible outcomes, growth, economic results and public recognition while mobilising, activating and including a wider circle of stakeholders. Hence here appears the co-creation of a diverse paradigm of entrepreneurship and driving values.

Conclusions

Acknowledging that the research findings are case study specific, the broader theoretical contribution of this study resides in prompting a more profound reflection on co-creation as an innovative and interdisciplinary process of collaboratively developing more complete, and deeper, levels of understanding. Without denying the potential of co-creation in terms of active participation of end-users as sources of product and service innovation (Von Hippel, 1989; Voorberg et al., 2015), this study focuses on how co-creation approaches help to develop and explore new knowledge by encouraging a process of thinking in an interdisciplinary, participative, 'out of the box' way. This has applications, and implications, for tourism, as this paper has shown, but also more generally, including indeed for the process of knowledge creation as an interdisciplinary exercise in academia.

Specifically, however, this study has revealed the ways in which small values-based tourism firms – through the voice of their owner-managers – critically reflect on the global economy and contemporary society sheds light on a diverse understanding of development, growth and global citizenship, and the associated implications for tourism. Using co-creation as a lens of investigation, the research findings convey a novel comprehension and re-interpretation of values that are – at the same time – different from profit-oriented neo-liberal paradigms, durable, and tied to a moral and ethical dimension of the identity construction process. The study discloses a realm of firms narrating themselves as proactive subjects within a challenging society and global economic system in which they co-create a diverse vision of tourism and values. Small values-based tourism firms aim to operate within a networked and empathic world permeated by a commonality of values, beliefs and goals. This is the co-creational dimension of such firms, constructing diverse values and paradigms together with a network of collaborators, vendors and customers, all presented firstly as proactive global citizens rather than merely stakeholders and customers.

The novelty of the study lies in its demonstration of a diverse understanding of small values-based tourism firms that are not seeking just to pursue the personal lifestyle goals of their founders (Ateljevic & Doorne, 2000), but are willing to activate and mobilise a global network of people – collaborators, providers and customers – towards new paradigms of development, growth and citizenship. These findings challenge a number of assumptions about small tourism firms that are not commercially oriented, while also offering novel theoretical and practical contributions. Small values-based tourism firms appear to reject the idea – as put forward by some authors – that they lack managerial and entrepreneurial skills (Ateljevic & Doorne, 2000; Gray, 2002; Hollick & Braun, 2005; Page, Forer, & Lawton, 1999; Sadler-Smith, Hampson, Chaston, & Badger, 2003; Skokic & Morrison, 2011). On the contrary, these firms attribute to themselves a diverse, extensive knowledge and knowhow going beyond the mere technicalities of a process of production-consumption and a quantitative profit-oriented paradigm. Participants picture a fully-rounded and inclusive model of development, growth and global citizenship embracing a diverse disruptive entrepreneurship, a tailor-made artisanal dimension, and an empathic, inclusive, transparent, dialectical, 'truth-telling' approach.

Small values-based tourism firms conceive 'smallness' and 'de-growth' not as a rejection of growth and business opportunities but rather as a way to control the firm's outcomes and quality through a selective approach valuing human relationships, inclusiveness, participation and conviviality. Participants' narratives frame their business either in terms of economic, profit-driven growth or in terms of a natural, value-driven growth guaranteeing independence and freedom in pursuing common goals. Their a-growth appeared to be functional in terms of reinforcing their identity, vision and mission against the failures in the global economy (Bauman, 2000, 2011; Latouche, 2004). Participants attributed to themselves managerial skills and an innovative co-creation culture framed not only in a proactive entrepreneurial mind-set but also in the idea of being citizens of a networked society first and owner-managers of businesses second.

Through the analysis of the first-person accounts of the owner-managers of small values-based tourism firms, the study sheds light on these firms' co-creation of alternative values and paradigms by revealing a connection between the narration of their private dimension as proactive subjects engaged in the debate on global issues, their disruptive entrepreneurial dimension, and the narration of their capacity to activate and include their stakeholders in their co-creative reasoning.

Disclosure statement

No potential conflict of interest was reported by the author.

References

Ateljevic, I., & Doorne, S. (2000). 'Staying within the fence': Lifestyle entrepreneurship in tourism. *Journal of Sustainable Tourism, 8*(5), 378–392.

Bauman, Z. (1998). *Globalization: The human consequences*. New York: Columbia University Press.

Bauman, Z. (2000). *Liquid modernity*. Cambridge, UK: Polity Press.

Bauman, Z. (2011). *Collateral damage: Social inequalities in a global age*. Cambridge, UK: Polity Press.

Benjamin, W., Jennings, M. W., & Doherty, B. (2008). *The work of art in the age of its technological reproducibility, and other writings on media*. London: Harvard University Press.

Butcher, J. (2008). 'Ethical' travel and well-being: Reposing the issue. *Tourism Recreation Research, 33*(2), 219–222.

Butler, R. (2011). Small is beautiful, but size can be important. *Tourism Recreation Research, 36*(2), 190–192.

Carless, D., & Sparkes, A. C. (2008). The physical activity experiences of men with serious mental illness: Three short stories. *Psychology of Sport and Exercise, 9*(2), 191–210.

Chase, S. E. (2005). Narrative inquiry: Multiple lenses, approaches, voices. In N. K. Denzin & Y. S. Lincoln (Eds.), *The Sage handbook of qualitative research* (3rd ed., pp. 651–679). Thousand Oaks, CA: Sage Publications Ltd.

Colclough, D. (1999). *Parrhesia*: The Rhetoric of free speech in early modern England. *Rhetorica: A Journal of the History of Rhetoric, 17*(2), 177–212.

Dewhurst, P., & Horobin, H. (1998). Small business owners. In R. Thomas (Ed.), *The management of small tourism and hospitality firms* (pp. 19–38). London: Cassell.

Easterly, W. (2006). *The white man's burden: Why the West's efforts to aid the rest have done so much ill and so little good*. New York: The Penguin Press.

Elkington, J. (1998). Cannibals with Forks: The triple bottom line of 21st century business (The Conscientious Commerce Series). Gabriola Island: New Society Publisher.

Foucault, M. (1983, October–November). Discourse and truth-the problematization of parrhesia. *6 lectures given by Michel Foucault at the University of California at Berkeley*. Retrieved from https://foucault.info/parrhesia/

Franklin, A. (2003). The tourist syndrome an interview with Zygmunt Bauman. *Tourist Studies, 3*(2), 205–217.

Gee, J. P. (1991). A linguistic approach to narrative. *Journal of Narrative and Life History, 1*(1), 15–39.

Gray, C. (2002). Entrepreneurship, resistance to change and growth in small firms. *Journal of Small Business and Enterprise Development, 9*(1), 61–72.

Hall, H. (2011). Consumerism, tourism and voluntary simplicity: We all have to consume, but do we really have to travel so much to be happy? *Tourism Recreation Research, 36*(3), 298–303.

Harrison, D. (2011). Tourism: Is small beautiful? *Tourism Recreation Research, 36*(2), 181–185.

Hollick, M., & Braun, P. (2005) Lifestyle entrepreneurship: The unusual nature of the tourism entrepreneur. In *Proceedings of the second Annual AGSE international entrepreneurship research exchange* (pp. 10–11). Melbourne: Swinburne Press.

Illich, I. (1971). *Deschooling society*. New York: Harper & Row.

Illich, I. (1973). *Tools for conviviality*. New York: Harper & Row.

Illich, I. (1997). The wisdom of Leopold Kohr. *Bulletin of Science, Technology & Society, 17*(4), 157–165.

Ind, N., & Coates, N. (2013). The meanings of co-creation. *European Business Review, 25*(1), 86–95.

Johansson, A. W. (2004). Narrating the entrepreneur. *International Small Business Journal, 22*(3), 273–293.

Klein, N. (2000). *No logo: Taking aim at the brand name bullies*. Toronto: Knopf Canada.

Klein, N. (2017). *No is not enough: Resisting Trump's shock politics and winning the world we need*. London: Haymarket Books.

Kohr, L. (1986). *The breakdown of nations*. London: Routledge.

Labov, W. (2010). Oral narratives of personal experience. In P. C. Hogan (Ed.), *Cambridge encyclopedia of the language sciences* (pp. 546–548). Cambridge: Cambridge University Press.

Labov, W., & Waletzky, J. (1997). Narrative analysis: Oral versions of personal experience. *Journal of Narrative & Life History, 7* (1–4), 3–38.

Lardinois, A. (2011). The parrhesia of young female choruses in ancient Greece. In L. Athenassaki, & E. Bowie (Eds.), *Archaic and classical choral song: Performance politics and dissemination* (pp. 161–172). Berlin: De Gruyter.

Latouche, S. (1993). *In the wake of the affluent society: An exploration of post-development*. London: Zed Books.

Latouche, S. (2004). Degrowth economics. *Le Monde Diplomatique*, 11. Retrieved from https://mondediplo.com/2004/11/14latouche

Lee, R. L. M. (2005). Bauman, liquid modernity and dilemmas of development. *Thesis Eleven*, *83*(1), 61–77.

MacAskill, W. (2015). *Doing good better: How effective altruism can help you make a difference*. New York, NY: Gotham.

Magala, S. (2009). *The management of meaning in organizations*. New York: Springer.

Merz, M. A., He, Y., & Vargo, S. L. (2009). The evolving brand logic: A service-dominant logic perspective. *Journal of the Academy of Marketing Science*, *37*(3), 328–344.

Novak, D. R. (2006). Engaging parrhesia in a democracy: Malcolm X as a truth-teller. *Southern Communication Journal*, *71*(1), 25–43.

Page, S. J., Forer, P., & Lawton, G. R. (1999). Small business development and tourism: Terra incognita? *Tourism Management*, *20*(4), 435–459.

Prahalad, C. K., & Ramaswamy, V. (2000). Co-opting customer competence. *Harvard Business Review*, *78*(1), 79–90.

Ramaswamy, V., & Gouillart, F. J. (2010). *The power of co-creation: Build it with them to boost growth, productivity, and profits*. New York: Simon and Schuster.

Rickly-Boyd, J. M. (2012). Authenticity & aura: A Benjaminian approach to tourism. *Annals of Tourism Research*, *39*(1), 269–289.

Riessman, C. K. (1993). *Narrative analysis*. Newbury Park, CA: Sage.

Ryan, B. (2012). Co-production: Option or obligation? *Australian Journal of Public Administration*, *71*(3), 314–324.

Sadler-Smith, E., Hampson, Y., Chaston, I., & Badger, B. (2003). Managerial behavior, entrepreneurial style, and small firm performance. *Journal of Small Business Management*, *41*(1), 47–67.

Sampaio, A. R., Thomas, R., & Font, X. (2011a). Small business management and environmental engagement. *Journal of Sustainable Tourism*, *20*(2), 179–193.

Schumacher, E. (1973). *Small is beautiful: Economics as though people mattered*. New York: Harper Colophon, Harper and Row.

Schumacher, E. F., & McKibben, B. (2010). *Small is beautiful: Economics as if people mattered*. New York: Harper Perennial.

Schumpeter, J. A. (1947). The creative response in economic history. *The Journal of Economic History*, *7*(02), 149–159.

Schumpeter, J. A. (2000). Entrepreneurship as innovation. In R. Swedberg (Ed.), *Entrepreneurship: The social science view* (Vol. 1, pp. 51–75). New York: Oxford University Press.

Sharpe, M. (2007). A question of two truths? Remarks on parrhesia and the 'political-philosophical' difference. *Parrhesia*, *2*, 89–108.

Shaw, G., & Williams, A. M. (2004). From lifestyle consumption to lifestyle production: Changing patterns of tourism entrepreneurship. In R. Thomas (Ed.), *Small firms in tourism: International perspectives (Advances in tourism research)* (pp. 99–113). Oxford: Elsevier.

Singer, P. (2010). *The life you can save: How to do your part to end world poverty*. New York: Random House Inc.

Singer, P. (2015). *The most good you can do: How effective altruism is changing ideas about living ethically*. New Haven: Yale University Press.

Singh, T. V. (Ed.). (2012). *Critical debates in tourism*. Bristol: Channel View Publications.

Skokic, V., & Morrison, A. (2011). Conceptions of tourism lifestyle entrepreneurship: Transition economy context. *Tourism Planning & Development*, *8*(2), 157–169.

Sluiter, I., & Rosen, R. M. (2004). *Free speech in classical antiquity: [Penn-Leiden Colloquium on ancient values, June 2002 at the University of Pennsylvania]*. Brill.

Smith, B., & Sparkes, A. C. (2008). Narrative and its potential contribution to disability studies. *Disability & Society*, *23*(1), 17–28.

Thomas, R., Shaw, G., & Page, S. J. (2011). Understanding small firms in tourism: A perspective on research trends and challenges. *Tourism Management*, *32*(5), 963–976.

Von Hippel, E. (1989). Cooperation between rivals: Informal know-how trading. In *Industrial dynamics* (pp. 157–175). Boston: Kluwer Academic Publishers.

Voorberg, W. H., Bekkers, V. J., & Tummers, L. G. (2015). A systematic review of co-creation and co-production: Embarking on the social innovation journey. *Public Management Review*, *17*(9), 1333–1357.

Weaver, D. (2011). Small can be beautiful, but big can be beautiful too—and complementary: Towards mass/alternative tourism synergy. *Tourism Recreation Research*, *36*(2), 186–189.

The changing face of the tour guide: one-way communicator to choreographer to co-creator of the tourist experience

Betty Weiler and Rosemary Black

Thirty years after Cohen's seminal work on tour guiding, the role(s) played by and skills required of tour guides continue to evolve. As 'experience' has come to be considered central to tourism, research on the guide as communicator and experience-broker has expanded. Guides broker experience in at least four domains – physical access, understanding, encounters and empathy. This conceptual paper examines, via the literature particularly on the mediatory and brokering roles of the tour guide and its intersections with social, economic and political trends, how and why the guide's role is changing. Together these bodies of literature on guiding and on societal trends are used to underpin a typology of future guided tour experiences distinguished by the target market, style of guiding and use of communication, with varying outcomes for the tourist. To meet the needs and expectations of twenty-first century tourists and the challenges of the global communication environment, tour guides need to become more highly skilled experience-brokers, including embracing technology to choreograph memorable experiences. To satisfy tourists in search of personalized and meaningful experiences, guides in some cases need to actively engage tourists in the co-creation of their own guided tour experiences. The typology provides a management and research framework for examining these relationships and their consequences.

Introduction

In the late 1970s, Schmidt (1979, p. 441) defined a guided tour as 'a form of tourism where the itinerary is fixed and known beforehand, and [involving] some form of planning and direct participation by agents apart from the tourists themselves'. As such, a tour guide has been defined as a person, usually a professional, who guides groups (and sometimes individuals) around venues or places of interest such as natural areas, historic buildings and sites, and landscapes, interpreting the cultural and natural heritage in an inspiring and entertaining manner (adapted from European Federation of Tour Guides Associations, 1998). Subsequent research on tour guiding has gone some way towards illuminating what tour guides do and the range of skills they need, including communication and interpretation.

That said, the emphasis in the above definition remains on one-way communication, positioning the *guide as presenter and entertainer* and the tour group as the audience. In the meantime, much has changed over the past three decades in our understanding of the tourist experience and, as will be illustrated later in the paper, in the context in which guided tours are now developed, marketed and delivered. Prior to the early 1970s the tourism literature was largely focused on tourism destinations and products. MacCannell's (1976) and Cohen's (1979) work launched decades of exploration and debate among scholars regarding the tourist experience as a construct, but it was not until Pine and Gilmore (1999) published *The Experience Economy*, arguing that memorable experiences are the key to competitive advantage and sustainability of any service provider, that the construct of experience was widely embraced as central to tourism. Pine and Gilmore (1999) posited that businesses should be perceived as 'theatres' and service personnel as 'performing artists'. This new perspective is evident in the 'performance turn' in the tourism literature where, like Pine and Gilmore, dramaturgical metaphors are used to describe how tourism occurs through encounters and enactments (Ek, Larsen, Hornskov, & Mansfeldt, 2008; Pearce, 2011). According to Pearce (2011, p. 3), the contributing components to the tourist experience as it is now understood are 'the sensory inputs, the affective reactions, the cognitive abilities to react to and understand the setting, the actions undertaken and the relevant relationships which define the participants' world'. Pearce likens the nature of the tourist experience to music produced by an orchestra where there are multiple sections each of which has its own elements, contributing different components at different times.

More recently, some scholars have observed and even advocated that the tourist be empowered to take a more active role as a creative, interactive agent (Richards & Wilson, 2006) and co-creator of tourist spaces (Ek et al., 2008; Mossberg, 2007). Ek et al. (2008) extended the dramaturgical metaphor to include not only service personnel but also tourists as performers who actively, corporeally, technically and socially perform and produce experiences, while Richards and Raymond (2000) argued that consumers are increasingly looking for more engaging, interactive experiences which can help them personalize and create meaning.

In summary, the tourism literature now widely embraces experience-centred tourism (Uriely, 2005) and describes tourist experiences as not only entertaining and memorable (Oh, Fiore, & Jeoung, 2007; Tung & Ritchie, 2011), but in some cases interactive and personal phenomena highly influenced by individual consumers seeking to create meaning (McIntosh & Siggs, 2005; Uriely, 2005). This suggests the need for a fresh look at guided tours and at the communicative role of the guide; traditional one-way communication appears too limiting in an experience- and consumer-centred tourism industry.

This paper is based on a systematic review of 280 research publications on tour guides and tour guiding, most having been published post-1990 (Weiler & Black, 2015), and in particular on the subset of 146 empirical studies and conceptual papers identified in scholarly journals. Several authors of this body of tour guiding literature have noted the need for a shift in tour guiding communication from presenter/entertainer to an experience focus, and have described the need to view tour group members as active participants and the tour guide as orchestrating or choreographing the experience (Edensor, 1998, as cited by Beedie, 2003). This paper assesses the capacity of guides to do this by drawing mainly on the body of research on *guides as experience-brokers*. Largely focused on cultural brokering and social mediation and consisting of 28 research papers, most of which were published in scholarly journals, this literature is the point of departure for the present paper. We first review this research on the mediatory/brokering roles of the guide, and also draw on key literature on visitor satisfaction with guides and guided tours to highlight the domains in which guides are required to broker experiences. We then present a selection of trends, particularly technological and socio-demographic trends that are shaping and potentially threatening the value and sustainability of guided tours, as a basis for arguing that guides may need to re-invent themselves if they are to continue to be relevant in the twenty-first century and add value to the tourist experience. The capacity of guides to adapt their communication approach and skills in response to these trends is central to their future and that of the guided tour itself.

The guide as experience-broker

For their recent book *Tour Guiding Research: Insights, Issues and Implications*, Weiler and Black (2015) used search engines and library databases to identify peer-reviewed journal papers, book chapters, conference papers, textbooks, doctoral-level theses and some grey literature on tour guiding. A critical review of this literature uncovered a body of research on mediation/brokering (e.g. Cohen, 1985; Dahles, 2002; Holloway, 1981; Jensen, 2010; Macdonald, 2006; Scherle & Nonnenmann, 2008; Weiler & Yu, 2007; for others, see Weiler & Black, 2015), as well as on visitors' satisfaction with guided tours (e.g. Arnould & Price, 1993; Ballantyne, Packer, & Sutherland, 2011; Bowie & Chang, 2005; Chang, 2006; Holloway, 1981; Huang, Hsu, & Chan, 2010; Hughes, 1991; Wolf, Stricker, & Hagenloh, 2013; Wong, 2001; for others, see Weiler & Black, 2015), that together provide a platform from which to examine how the guide's brokering role is and could be used to facilitate memorable experiences for tour participants. In the present paper, the terms 'mediator' and 'broker' are used interchangeably to include any active attempts by a guide to broker the experience, and may range from decisions about where and when to access a site, community or destination, to verbal communication, to role-modelling, to intervening and controlling what tourists see, hear and do. In the scholarly literature on tour guiding, the focus to date has been on mediation between visitors and destinations (host communities and environments) rather than on mediation within tour groups or indeed mediation of an individual's inner journey or experience, a point to which we return later in the paper.

It is worth mentioning that most, if not all, tourist experiences are highly mediated by both personal (human) and non-personal (often self-directed) media (Jennings & Weiler, 2006) including but not limited to guidebooks, websites, signs, visitor centres, podcasts and social media pages. In addition to tour guides who are formally employed to act as mediators/brokers, there are many others who formally and informally mediate, such as taxi and bus drivers, local residents and friends and relatives.

The body of research on tour guides' mediation and brokering does point to some distinct dimensions or domains of experience-brokering. Macdonald (2006) distinguished between *communicative* mediation (influencing how tourists understand a site or destination) and *interactional* mediation (influencing how, where, when and with whom tourists interact with host people and environments). Weiler and Yu (2007, p. 15) categorized cultural mediation into three domains or areas of influence: mediating physical *access*, mediating cognitive/affective access or *understanding* via the provision of information (roughly equivalent to Macdonald's communicative mediation), and mediating social access by facilitating the opportunity for *encounters* (roughly equivalent to Macdonald's interactional mediation). Finally, McGrath (2007) highlighted *emotional access* (empathy for host people/communities/cultures) as a fourth area that tour guides can influence as part of mediation, including empathy for historical people and cultures.

These researchers collectively investigated guiding in a range of mainly intercultural or heritage guiding contexts (Weiler & Black, 2015). Several of the researchers observed that mediation can both facilitate and inhibit tourist access and thus can contribute to a positive but also sometimes a negative experience across these domains.

While all empirical studies on mediation and brokering have limitations, collectively they illustrate the full range of mediatory roles played by the guide, particularly vis-à-vis the destination. Drawing on all of these schemas, Weiler and Black (2015) suggest the following framework to examine mediation by tour guides: (1) brokering physical access; (2) brokering encounters (interactions); (3) brokering understanding (intellectual access); and (4) brokering empathy (emotional access). The following sections highlight where research on mediation/brokering has focused on each of these domains of influence. As has already been suggested, aspects of mediation and brokering that are not evident in the literature such as tourist self-development and within-tour mediation are revisited later in the paper.

Domain 1: brokering physical access

Macdonald (2006) and Weiler and Yu (2007) acknowledged that guides broker physical access to places and spaces, while Arnould, Price, and Tierney (1998, p. 94) outlined how guides provide a 'cocoon of civilization' through which adventure tourists can experience the wilderness. Beedie (2003) described mountain guides as gatekeepers with technical and logistical know-how upon which adventure tourists depend for safety. Guides are important in staging the physical experience, including channelling and controlling tourists to be in the right place at the right time. They do this not only by physically manoeuvring tour groups but also by what Arnould et al. (1998) refer to as communicative staging; that is, by controlling what and how they present and interpret to tourists. In what MacCannell (1973) called staged authenticity, guides can focus on the 'front stage', they can introduce tourists to real and authentic back stages in response to tourists' desire for authenticity or they can construct a pseudo 'back stage' that gives tourists the impression of authenticity. In other words, tour guides can mediate physical access by not only facilitating what tourists see and experience but also by determining what is not revealed to tourists (Holloway, 1981).

In a cross-cultural context, Howard, Thwaites, and Smith (2001) found that Indigenous tour guides at an Australian national park mediated (limited) physical access to sites through the use of both communication and role-modelling. Guides can also broker physical access by providing tourists with opportunities to use all of their senses to appreciate the host culture and share and experience local stories, music and food (Weiler & Yu, 2007).

Domain 2: brokering encounters

As originally highlighted by Cohen (1985), tour guides can play an important role in mediating social interaction between tour group members and host communities. This has its challenges, particularly when there are cultural, social and economic disparities between tour group members, the guide and the host community. Both Macdonald (2006) and Weiler and Yu (2007) stressed that mediation involves brokering interactions between the group and host communities, for example, by providing language translation and facilitating two-way communication. Guides can also limit tourists' interactions with local people by drawing a group's attention inwards towards the guide and the tour group rather than outwardly directing it to the host community (Cohen, 1985; Holloway, 1981). Tour guides can passively or actively mediate external encounters between tourists and host communities and between tourists and staff working in hotels and tourist attractions. There is evidence of local guides in particular acting as a go-between and language broker, and as role models for appropriate environmental, social and cultural behaviour (Gurung, Simmons, & Devlin, 1996; Jensen, 2010; Ormsby & Mannle, 2006; Salazar, 2006; Weiler & Yu, 2007), although all of this is often difficult for guides who are not 'local' with sufficient understanding of and connections in the host environment.

While Cohen (1985) acknowledged that guides can also mediate interaction among tour group members, there has been little research focusing on the *social* mediatory role of the guide (compared to the cultural mediatory role) (Jensen, 2010; Macdonald, 2006) and no research has focused on the guide's role in mediating *within* tour groups (Weiler & Black, 2015). The latter is thus not explored in the present paper but is highlighted as an avenue for future research.

Domain 3: brokering understanding

The most researched domain of mediation is the guide as a broker of understanding (Macdonald, 2006; Weiler & Yu, 2007) or intellectual access (McGrath, 2007). Tour guides are considered by some to be the quintessential intercultural mediators of the tourism industry (Scherle & Kung, 2010; Scherle & Nonnenmann, 2008). In the former paper, the authors refer to the guides' informal and friendly communication with local people and their interpretation of local behaviours and customs as brokering cultural understanding. Tour guides can mediate understanding by using information and enrichment as a tool for conveying the significance of a place or site (Ap & Wong, 2001; Bras, 2000; Hughes, 1991). Often their multi-lingual skills bridge communication gaps. Many of the techniques used by guides to foster understanding and appreciation are well-known interpretive techniques such as non-verbal communication, asking questions, making use of anecdotes, examples, analogies and personal references, and using props (e.g. artefacts and photos from the past).

Similarly, Howard et al. (2001) concluded in their case study of Indigenous guides that guides mediated access to information (understanding) through the use of interpretive techniques and role-modelling, but also by challenging stereotypes and tourists' misconceptions about Australian Aboriginal culture. The guides' cultural brokering role revolved largely around communicating and interpreting local cultural values, both those of the site and those of Aboriginal society more generally.

There are also studies that report guides as potentially negative mediators, in the sense of inhibiting rather than fostering understanding. In Yogyakarta, Indonesia, Dahles (2002) found little evidence of the use of cultural mediation by guides to enhance the tourist experience. She attributed this to the political regime at the time, leading to tour guides being trained and directed to present a specific political commentary to tourists that promoted the government's views. The guides' performances were thus both staged and routinized, potentially limiting tourists' understanding. A similar conclusion was drawn by Mitchell's (1996) ethnographic case study of Malta. In an in-depth case study that analyzed the training and practice of tour guides, Mitchell demonstrated the critical role played by guides in creating and perpetuating positive images of a modern and highly European Malta and Maltese culture, expressly to support the country's application for accession into the European Union (EU). This is another example of a tourist's understanding of Malta's heritage and culture being highly mediated by the guides. A final example of potentially negative mediation of understanding is found in Mcgrath's (2007) critique of tour guiding of archaeological sites in Peru. She noted that the predominant type of guiding is the transfer of knowledge rather than cultural mediation as understood in a Western context.

While acknowledging that much of the research that underpins contemporary approaches to tour guide communication and interpretation has been in Western contexts, a body of research is emerging about effective and appropriate communication in other contexts such as Chinese culture (Xu, Cui, Ballantyne, & Packer, 2012). These studies reveal that the application of interpretation principles can be but are not always appropriate for experience-brokering in other cultures and may require adaptation.

Domain 4: brokering empathy

The domain of mediation that is least developed conceptually is that of empathy or emotion. Macdonald (2006) provided a useful historical review of early research focused on the role of cultural broker, highlighting a gulf of understanding and a certain lack of empathy between the culture of the visitor and the visited. Generally the term 'broker' implies action that bridges this divide, mainly as a means to enhance the experience of the tourist. McGrath suggests that guides need to help visitors 'get under the skin' (2007, p. 376) of visited areas rather than just providing physical and intellectual access. This can include brokering empathy and affinity not only for people and cultures but also for places and landscapes, by 'setting up extraordinary aesthetic experiences' (Beedie, 2003, p. 157). Modlin, Alderman, and Gentry (2011), in their case study of a southern US plantation house museum, provided an evocative illustration of how the emotional connection facilitated by museum guides can also lead to empathy for historical figures and sectors of society. They observed how guides use interpretive techniques such as storytelling to create not only cognitive but affective connections with the White American plantation owners. However, through selectivity in narrative content and in how and where they move visitors through the site they actually fail to create empathy for the enslaved community.

Although words such as script, narrative, storytelling, staging and interpretation are often used by these researchers in describing the techniques used by tour guides, Beedie (2003) is one of the few to refer specifically to the choreography of the guide. Drawing on the metaphor of a theatrical performance, he cited Edensor as having first developed the idea of the guide as choreographer in the context of tourists visiting the Taj Mahal (Edensor, 1998; as cited by Beedie, 2003). In the case of guides brokering the experience of climbing a mountain, Beedie (2003, p. 156) described how the guide 'choreographs the detail of the experience … where to walk, when to stop, how the group are positioned on and off rope, how to walk and conserve energy, how to move around obstacles … '. In addition to safety considerations, he argued that this is done to 'frame' parts of the experience, to 'reinforce communal solidarity', and 'to provide space for contemplation' (p. 157).

These bodies of research on experience-brokering illustrate that tour guides can and do use their communication skills to broker the experiences of their tour group, facilitating physical access, encounters, understanding and empathy on the part of their tour group. In fact, these examples of tour guides as experience-brokers illustrate the evolution and sophistication of the tour guide as a communicator, having moved well beyond the role of entertainer and one-way provider of information implied in definitions of the guide still used by international tourist guide associations (World Federation of Tourist Guide Associations, 2014). The literature on experience-brokering provides many concrete examples of guides using interpretation to choreograph experiences, not only through the telling of anecdotes and stories but also through the use of role-plays and drama (Davidson & Black, 2007; Skibins, Powell, & Stern, 2012), and in some cases through involving their clients as active participants in choreographing their own experience (Binkhorst & Den Dekker, 2009).

It is evident that much of a tour guide's experience-brokering relies on his/her verbal and non-verbal communication with tour groups and in some cases with host communities. However, many social, political and economic factors are influencing and challenging the traditional communication roles of the tour guide and the guided experience. The next section of the paper overviews some of these trends, identified as critical by interpretation and tour guiding experts in an online survey and a follow-up workshop conducted by the authors (Black & Weiler, 2013). The focus is particularly on information and communication technology and changing tourist profiles and preferences that are shaping the way guides may need to deliver their tours in the future. In particular, technological trends are challenging the extent and the way in which guides will be seen as (and will play a role) central to facilitating physical access, encounters, understanding and empathy both pre-departure and on-site. These developments include the widespread availability of the Internet, social media, mobile devices and other digital media. At the same time, the profile of the tourist is changing, becoming more diverse and more demanding. After examining some of these trends in the following section, we return to the issue of guides' communicative role(s) in order to broker memorable experiences and engage clients.

Trends and implications for tour guiding

The growth and availability of information and communication technology

The so-called digital revolution beginning with the explosion of the Internet / world-wide web has influenced and changed human communications and global productivity, providing a mass communication and social networking platform unprecedented in history (Bosco, 2006). Studies have shown a phenomenal penetration of the Internet globally (ABS, 2014; ITU, 2013; Pease, Rowe, & Cooper, 2007). For example, in Australia the number of households with access to the Internet at home continues to increase, representing 83% of all households in 2012–2013 (up from 6% in 1996) (ABS, 2014).

Following the Internet's inception as a mass provider of information via search engines such as Google, the appearance of other types of services have allowed for user-generated content via social media, including Wikipedia, YouTube, MySpace, Facebook and Twitter, that are reshaping the tourism industry (Buhalis, 2000; Pearce, 2011), creating a wide range of opportunities as well as threats for stakeholders, including tour guides. Digital media are used to learn about consumers; to provide them with information; for communication, transactions and relationships between various sectors of the tourism industry; and to influence tourist behaviours and experiences (Buhalis, 2000; Hannam, Butler, & Paris, 2014; Pease et al., 2007). From a tourist experience perspective, probably the most important advance of the past decade has been the development of user-friendly interfaces (Hannam et al., 2014) such as smart phones, tablets and MP3 players. The services enabled by these devices and web-based services have expanded communications including aps, podcasts, vodcasts, peer-to-peer networking, blogs, wikis, tag clouds and smart cards.

Buhalis (2000) highlighted many ways that digital technology enhances consumer satisfaction such as user-friendly and customized interfaces, more information and greater choice. These translate into greater involvement of consumers in planning their travel, building their own itineraries and managing their own experiences. Evidence suggests that, empowered by new tools such as mobile devices, consumers will increasingly seek and co-create personalized experiences (Anacleto, Figueiredo, Almeida, & Novais, 2014).

Over the past decade, electronic media and new technologies are also increasingly being used by the tourism industry to mediate and enrich the on-site tourist experience (Gretzel, Fesenmaier, Lee, & Tussyadian, 2011; Hannam et al., 2014; Kang & Gretzel, 2012; Wang, Park, & Fesenmaier, 2012), including mobile electronic 'tour guides' that are interactive and highly personalized such as via smart phones. Technology-assisted interpretation en route and on-site are becoming more common, including highly interactive visitor centres, phone aps, podcasts in the language of the visitor's choice (Kang & Gretzel, 2012) and Global Positioning System-triggered multi-media 'guided tours' (Chu, Lin, & Chang, 2012; Wolf et al., 2013). These devices allow the tourist to 'see backstage' and redistribute the power and control of staging and accessing experiences to the tourist. More than technological devices, they are social objects that allow tourists via social media to record and share their experiences and emotions, which in turn influences and informs others (Hannam et al., 2014; Jacobsen & Munar, 2012; Xiang & Gretzel, 2010).These technological changes and developments have implications for guided tours and tour guides' communication.

Socio-demographic trends: changing profiles and preferences of tourists

The volume of international travel is predicted to grow despite any slowdown in economies and growth in global conflict and risk. For example, in 1950, 25 million people took an international holiday; 100 years from now it is predicted to be 4.7 billion people (Yeoman, 2012).

While the overall volume of guided tours is also likely to grow, many socio-economic and demographic trends will potentially change the practice of tour guiding in the future. For example, indicators show that the world economic order is shifting, with Mexico, Brazil, Russia, India and China (MBRIC) emerging as the dominant economies (Stancil & Dadush, 2010; Yeoman, 2012). Personal prosperity is a key driver in modern society, thus growth in spending power of consumers from these emerging nations is driving growth in outbound tourism from these countries.

Paralleling this trend, Buhalis (2000) argued that personal prosperity has produced a 'new' technologically-savvy tourist who is a more knowledgeable, experienced and '… emboldened consumer-citizen, a more demanding, sophisticated and informed actor with intensified expectations of, for instance, quality innovation and premium choices, and efficient and ever-personalised customer service … ' (Yeoman, 2012, p. 23).

Demographic changes will also significantly impact on the nature of tourism demand. In most developed countries populations are ageing and there are falling birth rates and increased longevity (United Nations, 2014; Yeoman, 2012) with increased demand for well-being products and medical tourism (Richter, 2000). Up until 2030, most of the growth in tourism will be influenced by baby boomers retiring with their wealth; however, post-2030 it is predicted that retirees will lack sufficient funds to travel, especially in Europe where there will be pension policy reforms (Yeoman, 2012). Generation Y, that is, those born between 1977 and 2003, may form a significant tourist market. By 2020, this generation may well be the leaders, managers and consumers of travel experiences. A study of over 8000 Gen Y travellers cited by Benckendorff, Moscardo, and Pendergast (2010) found that they were intrepid, travelling more often, exploring more destinations, spending more on travel, booking more on the Internet, and getting a lot out of their travel. A number of key societal events are likely to have influenced Gen Y, including the digital revolution, global terrorism and global financial uncertainty (Benckendorff et al., 2010), and yet the extent to which they want to engage the services of tour guides has not been examined by researchers.

There are many other trends that may well affect experience-brokering, such as the enhanced awareness and pursuit of health and wellness (Smith & Kelly, 2006; Voigt, Brown, & Howat, 2011) and of personal growth and self-development (Devereux & Carnegie, 2006; Saunders, 2013) as part of travel. Such trends suggest that the guide's brokering of the tourist's 'inner journey' (Picard, 2012) may emerge as an equally or more important role than brokering the outer journey. The implications of these and other trends are revisited in the conclusion as fruitful avenues for future research.

Travellers' expectations regarding social and environmental responsibility

Scholars have noted a growth in environmental and social conscience of some sectors of the travelling public. For example, Pearce (2011) observed evidence of greater concern and interest in the environment and host communities among some tourists, suggesting that these tourists will be increasingly concerned and interested in environmental issues and will demand social and environmental responsibility from the tourism industry. Yeoman (2012) went further, arguing that tourists seek visible corporate commitment to tackling the environmental and ethical problems of the day by the industry. Traditionally, it has been difficult for tourists to gain information on the environmental and social responsibility of tourism suppliers, let alone find ways to personally support a local project or cause (O'Brien & Ham, 2012), but increasingly tourism businesses recognize the demand for these types of experiences.

There has also been an increase in travel philanthropy and fund raising (O'Brien & Ham, 2012; Powell & Ham, 2008) that can create valuable opportunities to involve visitors in supporting local projects or causes. Travel philanthropy refers to the goodwill of travellers in supporting the welfare of humankind and the conservation of nature in their travel destinations. Some of today's travellers, particularly from developed countries, are conscious of the wealth and resource disparity between their own country and developing countries. Volunteering or contributing in some way allows some tourists to make meaning from their travels by giving back to communities. The growth in volunteer tourism is also an example of some tourists' concern about environmental and social issues in host communities (Broad, 2003), with many travel companies now offering holidays with a volunteering component or opportunities to donate to foundations set up by these companies (Intrepid Travel, 2014; World Expeditions, 2014).

Implications and responses by the guided tour industry

The selected trends reviewed in the foregoing sections, that is, the growth and sophistication of digital technology, the changing profile of the tourist, their desire for meaningful travel experiences and their expectations of environmental and social responsibility, are trends that have been highlighted by interpretation and guided tour experts as being critical for the future of tour guiding (Black & Weiler, 2013). In particular, these trends have the potential to influence how guides will interact and communicate in order to broker future tourist experiences.

As evidenced in the review of literature, many tour guides are already well-equipped for and actively engaged in brokering the tourist experience, particularly in terms of gaining physical access. However, technology may be increasingly viewed by some as a substitute for the directional and access services provided by a tour guide, as well as one-way delivery of information, commentary, site interpretation and language translation. Zatori (2013, p. 125) noted that 'modern digital and web-based technologies allow companies to eliminate the tour guide physically' for these types of services. For example, digitized guide books are now available through mobile applications designed for smart phone operators such as Nokia, Apple, Google and Android (Yeoman, 2012).

On the other hand, the widespread availability of guidebooks and electronic media for information gathering and navigation for tourists frees up tour guides to progress from the delivery of one-way commentary to a greater emphasis on experience-brokering, particularly the non-physical dimensions of brokering highlighted in the literature: brokering understanding, encounters and empathy. In some contexts, high-quality interpretation and personalized mediation may be the key reasons for employing a guide in the future (Mcgrath, 2007). This underlines the need for the guide's communication to be innovative and marketed as a service that adds value in ways that technology cannot.

Of course, technology can be embraced by guides to enhance the delivery of guided tours, such as incorporating podcasts and smart phone aps into their tours. A tour guide's communication can be interactive with both tourists and host communities, and technology also allows guides to customize the experience to tourists' needs and expectations. Based on Gen Y's high use of and familiarity with technology and desire for new experiences, it appears this market segment will be comfortable with and expect the integration of digital media into their travel experiences pre-, on-site and post-tour.

Evidence from the literature together with changing socio-economic and demographic trends suggests that future tourists will be more demanding and sophisticated and seek their guides to broker more engaging, interactive experiences that help them personalize and create meaning (Pine & Gilmore, 1999; Richards & Raymond, 2000). In the case of Gen Y tourists who are adventurous, enjoy travel and seek meaningful experiences, guides can play an important part in brokering empathy and affinity, not only for cultures and communities but also as change agents and facilitators of volunteer experiences.

With regard to tourists' expectations of environmental and social responsibility, guides may be engaged as caretakers and monitors of both tourists and their behaviour, and as information trackers in relation to tourism impacts. Similar to the use of volunteer tourists who engage in data collection for the purposes of monitoring impacts and changes in natural and cultural environments, guides can be trained, equipped and rewarded for such activities. Guided tours are already being used as a tool for change in urban areas for participants to visualize possible futures (Hansson, 2009). This type of guided experience is a form of citizen participation where people are being taken through the process of 'experiencing – understanding – taking action'. This is also happening with art tours where some guides are adopting the role of provocateur and change agent by presenting alternative view points and counter-culture narratives. In a similar vein, in Europe young people are employed as guide-activists, guiding and facilitating groups of young people through cities using experiential activities to raise tourists' awareness of environmental and social issues and to engage them in solutions (www.touristsavetheworld.com). Tour guides who can adapt their communication approach to foster environmentally and socially responsible outcomes may well be in greater demand and more highly valued by some tourist segments in the future than guides who simply deliver information or even those who broker interactive experiences that are primarily tourist-focused.

Whether or not assisted by technology, whether targeted at Gen Y or environmentally and socially aware tourists seeking to do something meaningful with their travel, tour guides can channel their communication expertise to personalize, customize and even co-create tours in ways that truly value-add to the tourist experience. The concept of co-creation is relatively new to the tour guiding literature, and is discussed more fully in the following section.

The guide as co-creator

As noted at the outset of this paper, much of the literature on tour guides as experience-brokers positions the guide as the choreographer and even the controller of the guided tour experience, with the tourist playing a more passive or at best a reactive role. However, as tourists gain experience and confidence, they may seek intellectual or emotional engagement both in the planning and execution of the guided experience and some may seek a sense of ownership or control. Rather than leaving these tourist segments to their own devices, tour guides can provide the space and opportunity for these types of experiences. Using the metaphor of a play, Prebensen and Foss (2011), like Beedie (2003), described tourists as actors and tour companies providing the set, other actors and scenes. These tourists have the capacity and expectation of actively contributing to the design and production of their own experiences (Binkhorst & Den Dekker, 2009; Richards & Wilson, 2006). In this context, Binkhorst and Den Dekker (2009) argued for a guiding approach that goes beyond choreography; if control is appropriately balanced based on the strength of the actors, the guide can facilitate self-expression and a sense of ownership by tourists, thereby allowing for co-creation of the experience (Binkhorst & Den Dekker, 2009).

As illustrated in Table 1, co-creation of guided tour experiences requires flexibility, commitment to transparency, openness to dialogue and investment of time on the part of the tour operator and the guide. In the first instance, the guide needs to invite tourists to actively contribute to constructing the experience and provide avenues for them to do so. This requires communication at an individual and group level before, during and even after the tour. Morgan, Elbe, and de Esteban Curiel (2009) noted that co-creation also requires a mindset by the tourism operator (as well as the guide and tourist) that the product (the tourist experience) will not always be the same, even for tourists on the same tour at the same time. As outlined in column two of Table 1, guides need to develop different communication skill sets involving listening and facilitating rather than presenting and entertaining, and this will require a degree of adaptability, creativity and the capacity to innovate that may be strengthened

Table 1. Co-creation and its implications for tour guides.

Co-creation experience concept: Adapted from Zatori (2013)	Implications for tour guides: Authors' application of co-creation to tour guides
• The essential building blocks of experience co-creation are dialogue, access and transparency. • Co-creation of experience requires interaction between the consumers and the organization and is in the organization's best interest. • Products and services are intermediaries of co-created experiences. • Organizations can provide an experience environment for the creation of personalized experiences; the product may be the same but customers can be encouraged to construct different experiences. • The individual consumer and his/her co-creation of the experience (with the company) are at the centre of the value creation process. • Consumers need not stand alone; they can form a consumer community and co-create an experience together. • To enhance experience co-creation with the consumer, organizations should co-operate and network. • The context and the level of consumer involvement will contribute to the sense of ownership, level of personalization and meaning that the customer takes away from co-creation.	• Guide offers to broker communication between customers and tour company. • Group members are encouraged and acknowledged by the guide for their contribution. • Guide needs to listen and foster open communication among all parties. • Guide needs to encourage and provide opportunities and examples of how group can value-add to aspects of tour 'product'. • The tour guide needs to communicate with tourists individually and as a group to identify expectations, preferably prior to the start of the tour. • Time needs to be allocated at different times for group discussion with the guide about all aspects of the tour experience. • Guide needs to provide time for the group to discuss the tour on their own. • Guide may seek resources outside the company to facilitate the experience. • Guide needs to use tools and strategies to encourage personal reflection by the group. • Guide uses communication during and post-tour to enhance memorability and meaningfulness of the tour.

Source: Adapted from Zatori (2013).

Table 2. A typology of guided tours, tour guiding communication and tourist outcomes.

Type of tour experience:	Examples of target markets:	How tour group is perceived by the guide:	How tour guide is perceived by the tour group:	Tour guide's communication style:	Potential for use of technology as part of the experience :	Communication skills required:	Outcomes for tourists:
Traditional group tour	MBRIC* and other inexperienced travellers	Audience (passive and reactive)	Entertainer/ presenter • guide in control	*Commentary/ script*	To add value	Presentation skills (verbal and non-verbal)	Enjoyable
Experience-focused group tour	Experienced travellers, e.g. baby-boomers	Actors (passive or active)	Choreographer • guide in control	*Experience-brokering: especially encounters, understanding, and empathy*	To add value	Presentation skills (verbal and non-verbal) plus Interpretation and mediation skills	Enjoyable and memorable
Customized / personalised tour	Gen Y; Socially and environmentally responsible travellers	Co-creators (pro-active)	Co-creator • shared control	*Variable – customized to groups, individuals and contexts*	To add value and customize experiences	Presentation skills (verbal and non-verbal) plus Interpretation and mediation skills plus adaptability/ improvisation skills	Enjoyable and memorable and meaningful

Source: Authors; *Mexico, Brazil, Russia, India and China (Yeoman, 2012).

through guide recruitment and selection, together with training, education and professional development. This is an area that requires more attention by researchers, practitioners and educators/trainers.

The changing nature of tour guide communication: a typology

Drawing on the foregoing review of research on tour guides' roles in mediating experiences together with current trends, we now turn to mapping the factors that will influence how guides communicate with tourists in the future, against their communication styles, skills and outcomes. This is presented as a typology (Table 2), in recognition that there is not a 'one-size-fits-all' approach to tour guide communication and in the hope of providing a foundation for further research.

While a commentary or script-based style of communication may continue to suit some guiding contexts and clients, including first-time MBRIC travellers, there are at least two contexts in which a radically different style of communication is needed (see Table 2). The first context is an *experience-focused group tour*, where tourists may be either passive or active participants. In this type of tour, the guide choreographs the experience for the group, and uses verbal, non-verbal, interpretive and mediatory skills to broker physical access for the group, but more importantly to also broker understanding, encounters and empathy. The result is an experience that is both enjoyable and memorable. The second context is a *customized or personalized tour*, where tourists must be active participants. The guide, together with tourists, co-creates this type of tour, with the guide once again harnessing his/her communicative (particularly listening), interpretive and mediatory skills but also applying adaptability and improvisational skills. The guide uses these skills to broker the experience on a range of dimensions as shown in Table 2. The result for the tourist is an experience that is enjoyable and memorable but also meaningful, because the tourist has been able to channel the guide's communication towards his/her own needs and interests. In all tour contexts, information and communication technology can value-add to the tour.

The typology presented in Table 2 demonstrates that there is still room for the *traditional guided tour*, and that there is still a need for guides in all types of tours to communicate in ways that provide an enjoyable experience. However, the previous discussion on the changing nature of tourism, society and technology suggests that guides need to deepen and broaden their communication and experience-brokering if they are to remain relevant and appealing to the full range of potential clients. The literature on mediation provides a point of departure regarding the skills and techniques required to be a good experience-broker, particularly brokering access, encounters and understanding. The changes in digital technology reviewed in this paper provide challenges as well as opportunities for guide communication, and in particular underline the need for guides to embrace technology in order to add value to the tourist experience.

If the demand for co-created tours continues to grow, it is guides, prepared to adapt and improvise their communication, who will be best positioned to meet this demand. Mossberg (2007) suggested that professionals in creative industries, such as actors trained in performance, could be useful in roles as educators, entertainers and value creators, and this may well be true of future tour guides.

Conclusions

It is clear that the traditional communicative role of the tour guide as a one-way presenter and entertainer is inadequate for the marketplace of twenty-first century tourism. The literature points to the value of an experience-centred approach for improving and maintaining the relevance of guided tours, and thus to the need for guides to be equipped to deliver enjoyable, memorable and in some cases interactive and personally-relevant experiences. Together with responding to technological and socio-demographic trends, guides need to utilize new and diverse communication approaches. The future of the tour guiding industry requires guides to choreograph and in some cases co-create experiences, which in turn requires guides to adapt their communication approach and skills to match the context and client expectations.

There are several avenues requiring more research in order to better advance the tour guide as an effective communicator in the twenty-first century. A nuanced understanding of the guide as choreographer, and particularly in relation to brokering empathy with host communities as well as relationship building within tour groups, is fundamental and in need of in-depth research. In addition, virtually no research has examined the guide as co-creator, as discussed in the previous section of the paper. The antecedents and outcomes of these two approaches to tour guiding as depicted in Table 2 need further exploration and testing.

In addition to the selected trends explored in this paper, there are others that may well impact tour guiding. For example, there is evidence that some tourists may look to their guide to facilitate or support a personal inner journey towards self-development. All guides need to be cognizant that, especially for longer tours, the journey for the individual participant may be an inner one as well as an outer one. While the roles of the guide in brokering dimensions of the outer experience have been researched, the guide as broker or facilitator of the tourist's inner journey has been largely unexplored.

In some cases tourists may be searching for personal meaning, intellectual or spiritual enlightenment, or life change, and thus guides may need to develop skills as change agents. The outdoor education literature has engaged with the concept of personal transformation for many years (Mortlock, 2001) and may offer insights to researchers and to guides wishing to enhance their capacity to foster tourists' personal growth. It may well be that transformation can be an outcome for both experienced-focused tours and customized/personalized (co-created) tours. That said, not every tourist seeks to be transformed or changed and even a tour designed to be transformational may not achieve this for all tour group members.

In conclusion, trends such as changing technologies, socio-demographics and tourists' expectations come with the need for broadening how the communicative role of the guide is conceptualized. Many tourists want not only enjoyable but memorable and in some cases meaningful experiences, including the opportunity to co-create experiences that are customized and personalized to their interests and needs. Research that can elucidate this new approach to communication by tour guides is central if tour guides in the twenty-first century are to maintain relevance and guided tours are to be sustained.

Acknowledgement

The authors wish to acknowledge the valuable contributions of the anonymous reviewers of this paper which have enabled them to greatly improve the manuscript.

Disclosure statement

No potential conflict of interest was reported by the authors.

References

Anacleto, R., Figueiredo, L., Almeida, A., & Novais, P. (2014). Mobile application to provide personalised sightseeing tours. *Journal of Network and Computer Applications*, *41*, 56–64.

Ap, J., & Wong, K. K. F. (2001). Case study on tour guiding: Professionalism, issues and problems. *Tourism Management*, *22*, 551–563.

Arnould, E. J., & Price, L. L. (1993). River magic: Extraordinary experience and the extended service encounter. *Journal of Consumer Research*, *20*(1), 24–45.

Arnould, E. J., Price, L. L., & Tierney, P. (1998). Communicative staging of the wilderness servicescape. *Service Industries Journal*, *18*(3), 90–115.

Australian Bureau of Statistics. (2014). *Household use of information technology, Australia 2012–2013*. Retrieved from http://www.abs.gov.au/ausstats/abs@.nsf/Lookup/8146.0Chapter12012-13

Ballantyne, R., Packer, J., & Sutherland, L. (2011). Visitors' memories of wildlife tourism: Implications for the design of powerful interpretive experiences. *Tourism Management*, *32*, 770–779.

Beedie, P. (2003). Mountain guiding and adventure tourism: Reflections on the choreography of the experience. *Leisure Studies*, *22*, 147–167.

Benckendorff, P., Moscardo, G., & Pendergast, D. (Eds.). (2010). *Tourism and generation Y.* Wallingford, Oxon: CAB International.

Binkhorst, E., & Den Dekker, T. (2009). Agenda for co-creation tourism experience research. *Journal of Hospitality Marketing and Management*, *18*, 311–327.

Black, R., & Weiler, B. (2013). *Opportunities and threats to the practice of face-to-face interpretation in Australia*. Paper presented at Visitor Studies seminar, University of Canberra, February, 2013.

Bosco, J. (2006). *Tools, culture and communications: Past – present – future*. Sydney, Australia: Global Summit. Retrieved from http://www.educationau.edu.au/jahia/

webdav/site/myjahiasite/shared/globalsummit/JBosco_GS2006.pdf

Bowie, D., & Chang, J. C. (2005). Tourist satisfaction: A view from a mixed international guided package tour. *Journal of Vacation Marketing*, *11*, 303–322.

Bras, K. (2000). Dusun sade: Local tourist guides, the provincial government and the (re)presentation of a traditional village in Central Lombok. *Pacific Tourism Review*, *4*(2/3), 87–103.

Broad, S. (2003). Living the Thai life: A case study of volunteer tourism at the Gibbon rehabilitation project, Thailand. *Tourism Recreation Research*, *28* (3), 63–72.

Buhalis, D. (2000). Trends in information technology and tourism. In W. C. Gartner & D. W. Lime (Eds.), *Trends in outdoor recreation, leisure and tourism* (pp. 47–61). Wallingford, Oxon: CAB International.

Chang, J. C. (2006). Customer satisfaction with tour leaders' performance. A study of Taiwan's package tours. *Asia Pacific Journal of Tourism*, *11*, 97–116.

Chu, T., Lin, M., & Chang, C. (2012). mGuiding (mobile guiding) – Using a mobile GIS app for guiding. *Scandinavian Journal of Hospitality and Tourism*, *12*, 269–283.

Cohen, E. (1979). Rethinking the sociology of tourism. *Annals of Tourism Research*, *6*(1), 18–35.

Cohen, E. (1985). The tourist guide: The origins, structure and dynamics of a role. *Annals of Tourism Research*, *12* (1), 5–29.

Dahles, H. (2002). The politics of tour guiding: Image management in Indonesia. *Annals of Tourism Research*, *29*, 783–800.

Davidson, P., & Black, R. (2007). Voices from the profession: Principles of successful guided cave interpretation. *Journal of Interpretation Research*, *12* (2), 25–44.

Devereux, C., & Carnegie, E. (2006). Pilgrimage: Journeying beyond self. *Tourism Recreation Research*, *31*(1), 47–56.

Edensor, T. (1998). *Tourists at the Taj*. London: Routledge.

Ek, R., Larsen, J., Hornskov, S. B., & Mansfeldt, O. K. (2008). A dynamic framework of tourist experiences: Space-time and performances in the experience economy. *Scandinavian Journal of Hospitality and Tourism*, *8*, 122–140.

European Federation of Tour Guides Associations (ETFGA). (1998). *European federation of tour guides associations brochure*. Vienna, Austria: European Federation of Tour Guides Associations.

Gretzel, U. D., Fesenmaier, Y., Lee, J., & Tussyadian, I. (2011). Narrating travel experiences: The role of new media. In R. Sharpley & P. Stone (Eds.), *Touristic experiences: Contemporary perspectives* (pp. 171–82). New York: Routledge.

Gurung, G., Simmons, D., & Devlin, P. (1996). The evolving role of tourist guides: The Nepali experience. In R. Butler & T. Hinch (Eds.), *Tourism and indigenous peoples* (pp. 108–28). London: International Thomson Business Press.

Hannam, K., Butler, G., & Paris, C. M. (2014). Developments and key issues in tourism mobilities. *Annals of Tourism Research*, *44*, 171–185.

Hansson, E. (2009). Guided tours as tools for change – taking the planning dialogue into – and in – the streets. In P. Adolfsson, P. Dobers, & M. Jonasson (Eds.), *Guiding and guided tours* (pp. 63–84). Goteborg, Sweden: BAS Publishers.

Holloway, J. C. (1981). The guided tour a sociological approach. *Annals of Tourism Research*, *8*(3), 377–402.

Howard, J., Thwaites, R., & Smith, B. (2001). Investigating the roles of the indigenous tour guide. *Journal of Tourism Studies*, *12*(2), 32–39.

Huang, S., Hsu, C. H. C., & Chan, A. (2010). Tour guide performance and tourist satisfaction: A study of the package tours in Shanghai. *Journal of Hospitality and Tourism Research*, *34*(1), 3–33.

Hughes, K. (1991). Tourist satisfaction: A guided cultural tour in North Queensland. *Australian Psychologist*, *26*(3), 166–171.

International Telecommunication Union (Itu). (2013). *ITU measuring the information society*. Retrieved from http://www.itu.int/en/ITU-D/Statistics/Pages/publications/default.aspx

Intrepid Travel. (2014). *The Intrepid Foundation*. Retrieved from http://www.intrepidtravel.com/about-intrepid/intrepid-foundation

Jacobsen, J. K. S., & Munar, A. M. (2012). Tourist information search and destination choice in a digital age. *Tourism Management Perspectives*, *1*, 39–47.

Jennings, G. R., & Weiler, B. (2006). Mediating meaning: Perspectives on brokering quality tourism experiences. In G. R. Jennings & N. Nickerson (Eds.), *Quality tourism experiences* (pp. 57–78). Burlington, Massachusetts: Elsevier.

Jensen, O. (2010). Social mediation in remote developing world tourism locations: The significance of social ties between local guides and host communities in sustainable tourism development. *Journal of Sustainable Tourism*, *18*, 615–633.

Kang, M., & Gretzel, U. (2012). Effects of podcast tours on tourist experiences in a national park. *Tourism Management*, *33*, 440–455.

MacCannell, D. (1973). Staged authenticity: Arrangements of social space in tourist settings. *American Journal of Sociology*, *79*, 589–603.

MacCannell, D. (1976). *The tourist: A new theory of the leisure class*. London: Macmillan.

Macdonald, S. (2006). Mediating heritage: Tour guides at the former Nazi Party rally grounds, Nuremberg. *Tourist Studies*, *6*(2), 119–138.

McGrath, G. (2007). Towards developing tour guides as interpreters of cultural heritage: The case of Cusco, Peru. In R. Black & A. Crabtree (Eds.), *Quality assurance and certification in ecotourism* (pp. 364–94). Wallingford, Oxfordshire: CABI.

McIntosh, A. J., & Siggs, A. (2005). An exploration of the experiential nature of boutique accommodation. *Journal of Travel Research*, *44*(1), 74–81.

Mitchell, J. P. (1996). Presenting the past: Cultural tour-guides and the sustaining of European identity in Malta. In L. Briguglio, R. Butler, D. Harrison, & W. Leal Filho (Eds.), *Sustainable tourism in Islands and small states: Case studies* (pp. 199–219). London: Cassell PLC.

Modlin, E. A., Alderman, D. H., & Gentry, G. W. (2011). Tour guides as creators of empathy: The role of affective inequality in marginalizing the enslaved at plantation house museums. *Tourist Studies*, *11*(1), 3–19.

Morgan, M., Elbe, J., & de Esteban Curiel, J. (2009). Has the experience economy arrived? The views of destination managers in three visitor-dependent areas. *International Journal of Tourism Research*, *11*, 201–216.

Mortlock, C. (2001). *Beyond adventure: reflections from wilderness – an inner journey.* Milnthorpe, UK: Cicerone Press.

Mossberg, L. (2007). A marketing approach to the tourist experience. *Scandinavian Journal of Hospitality and Tourism*, *7*, 59–74.

O'Brien, T., & Ham, S. H. (2012). *Toward professionalism in tour guiding: A manual for trainers.* Washington, DC: US Agency for International Development.

Oh, H., Fiore, A., & Jeoung, M. (2007). Measuring experience economy concepts: Tourism applications. *Journal of Travel Research*, *46*(2), 119–132.

Ormsby, A., & Mannle, K. (2006). Ecotourism benefits and the role of local guides at Masoala National Park, Madagascar. *Journal of Sustainable Tourism*, *14*, 271–287.

Pearce, P. (2011). *Tourist behaviour in the contemporary world*. Bristol: Channel View Publications.

Pease, W., Rowe, M., & Cooper, M. (2007). *Information and communication technologies in support of the tourism industry.* Hershey, PA: Idea Group Publishing.

Picard, D. (2012). Tourism, awe and inner journeys. In D. Picard & M. Robinson (Eds.), *New directions in tourism analysis: Emotion in motion: Tourism, affect and transformation* (pp. 1–20). Farnham, Surrey, GBR: Ashgate Publishing Group.

Pine, B. J.Jr., & Gilmore, J. H. (1999). *The experience economy: Work is theatre and every business a stage.* Boston, MA: Harvard Business Review Press.

Powell, R. B., & Ham, S. S. (2008). Can ecotourism interpretation really lead to pro-conservation knowledge, attitudes and behaviour? Evidence from the Galapagos Islands. *Journal of Sustainable Tourism*, *16*, 467–489.

Prebensen, C. K., & Foss, L. (2011). Coping and co-creating in tourist experiences. *International Journal of Tourism Research*, *13*(1), 54–67.

Richards, G., & Raymond, C. (2000). Creative tourism. *ATLAS news*, 23, 16–20.

Richards, G., & Wilson, J. (2006). Developing creativity in tourist experiences: A solution to the serial reproduction of culture? *Tourism Management*, *25*, 297–305.

Richter, L. K. (2000). Tourism challenges in developing nations: Continuity and change in the millennium. In D. Harrison (Ed.), *Tourism in the less developed world: Issues and case studies* (pp. 47–59). Wallingford, Oxon: CAB International.

Salazar, N. B. (2006). Touristifying Tanzania: Local guides, global discourse. *Annals of Tourism Research*, *33*, 833–852.

Saunders, R. (2013). Identity, meaning and tourism on the Kokoda Trail. In A. Norman (Ed.), *Journeys and destinations: Studies in travel, identity, and meaning* (pp. 23–45). Newcastle upon Tyne: Cambridge Scholars Publishing.

Scherle, N., & Kung, H. (2010). *Cosmopolitans of the 21st Century? Conceptualising tour guides as intercultural mediators*. Paper presented at the First International Research Forum on Guided Tours, 23–25 April 2009, Halmstad University, Sweden.

Scherle, N., & Nonnenmann, A. (2008). Swimming in cultural flows: Conceptualising tour guides as intercultural mediators and cosmopolitans. *Journal of Tourism and Cultural Change*, *6*(2), 120–137.

Schmidt, C. J. (1979). The guided tour: Insulated adventure. *Urban Life*, *7*, 441–467.

Skibins, J. C., Powell, R. B., & Stern, M. J. (2012). Exploring empirical support for interpretation's best practices. *Journal of Interpretation Research*, *17*(1), 25–44.

Smith, M., & Kelly, C. (2006). Wellness tourism. *Tourism Recreation Research*, *31*(1), 1–4.

Stancil, B., & Dadush, U. (2010). The world order in 2050. *Carnegie Endowment for International Peace Policy Outlook*, April, 1–29.

Tung, V. W. S., & Ritchie, J. R. (2011). Exploring the essence of memorable tourism experiences. *Annals of Tourism Research*, *38*, 1367–1386.

United Nations. (2014). *Social indicators*. Retrieved from http://unstats.un.org/unsd/demographic/products/socind/

Uriely, N. (2005). The tourist experience: Conceptual developments. *Annals of Tourism Research*, *32*, 199–216.

Voigt, C., Brown, G., & Howat, G. (2011). Wellness tourists: In search of transformation. *Tourism Review*, *66*, 16–30.

Wang, D., Park, S., & Fesenmaier, D. R. (2012). The role of smartphones in mediating the touristic experience. *Journal of Travel Research*, *51*, 371–387.

Weiler, B., & Black, R. S. (2015). *Tour guiding research: Insights, issues and implications*. Bristol, UK: Channel View Publications.

Weiler, B., & Yu, X. (2007). Dimensions of cultural mediation in guiding Chinese tour groups: Implications for interpretation. *Tourism Recreation Research*, *32*(3), 13–22.

Wolf, I. D., Stricker, H. K., & Hagenloh, G. (2013). Interpretive media that attract park visitors and enhance their experiences: A comparison of modern and traditional tools using GPS tracking and GIS technology. *Tourism Management Perspectives*, *7*, 59–72.

Wong, A. (2001). Satisfaction with local tour guides in Hong Kong. *Pacific Tourism Review*, *5*(1), 59–67.

World Expeditions. (2014). The world expeditions foundation. Retrieved from http://www.worldexpeditions.com/au/index.php?section=about_usandid=122521

World Federation of Tourist Guide Associations (WFTGA). (2014). What is a tourist guide? Retrieved from http://www.wftga.org/tourist-guiding/what-tourist-guide

Xiang, Z., & Gretzel, U. (2010). Role of social media in online travel information search. *Tourism Management*, *31*, 179–188.

Xu, H. G., Cui, Q. M., Ballantyne, R., & Packer, J. (2012). Effective environmental interpretation at Chinese natural attractions: The need for an aesthetic approach. *Journal of Sustainable Tourism*, *21*, 117–133.

Yeoman, I. (2012). *2050 – Tomorrow's tourism*. Bristol, UK: Channel View Publications.

Zatori, A. (2013). The impact of the experience management perspective on tour providers. In D. Koerts & P. Smith (Eds.), *3rd international research forum on guided tours* (pp. 125–137). Breda, Netherlands: NHTV Breda University of Applied Sciences.

Index

Page numbers in **bold** refer to tables and those in *italic* refer to figures.